RAND SOCIAL AND ECONOMIC WELL-BEING

T0306636

Updating the Costs of Compliance for California's Hospital Seismic Safety Standards

Benjamin Lee Preston, Tom LaTourrette, James R. Broyles, R. J. Briggs, David Catt, Christopher Nelson, Jeanne S. Ringel, Daniel A. Waxman

Prepared for the California Hospital Association

For more information on this publication, visit www.rand.org/t/RR3059

Library of Congress Control Number: 2019939398

ISBN: 978-1-9774-0263-9

Published by the RAND Corporation, Santa Monica, Calif.

Preface

The 1994 Northridge earthquake was one of the more severe seismic events to affect California during the 20th century. As with previous events, including the 1989 Loma Prieta earthquake and the 1971 Sylmar earthquake, Northridge prompted a range of policy responses designed to further increase the resilience of California to future seismic events. The impact of the Northridge earthquake on hospitals was a particular concern. A reported 11 hospitals suffered structural damage or loss of electrical and water services. In some cases, the damage necessitated the evacuation of the hospital. This led to a legislative response in the form of Senate Bill (SB) 1953—an update to prior legislation that focused on upgrading or replacing older hospital buildings to enhance resilience to seismic events. SB 1953 established deadlines for ensuring the life safety of hospitals during seismic events, as well as for ensuring continued operations postevent. Prior studies published in 2002 and 2007 provided early insights into the implementation of SB 1953 and its potential costs.[1] However, in the time that has elapsed since the most recent study, much progress has been made toward compliance.

We frame this update on SB 1953 implementation around three core analyses: forward-looking costs of compliance, the affordability of the legislation for hospitals, and the exploration of policy alternatives that could reshape how SB 1953 is implemented in the future. In addition, we identify the various factors that influence how hospitals prioritize compliance strategies and how SB 1953 compares with other policies designed to increase the resilience of critical infrastructure. Our intended audience includes hospital management and executives, California policymakers, and the communities throughout the state that depend on robust and reliable health care service delivery.

This project was funded by the California Hospital Association.

RAND Social and Economic Well-Being

RAND Social and Economic Well-Being is a division of the RAND Corporation that seeks to actively improve the health and social and economic well-being of populations and communities throughout the world. This research was conducted in the Community Health and Environmental Policy Program within RAND Social and Economic Well-Being. The program focuses on such topics as infrastructure, science and technology, community design, community health promotion, migration and population dynamics, transportation, energy, and climate and the environment, as well as other policy concerns that are influenced by the natural and built

[1] Charles Meade, Jonathan Kulick, and Richard Hillestad, *Estimating the Compliance Costs for California SB 1953*, Oakland: California HealthCare Foundation, 2002; Charles Meade and Jonathan Kulick, *SB1953 and the Challenge of Hospital Seismic Safety in California*, Oakland: California HealthCare Foundation, 2007.

environment, technology, and community organizations and institutions that affect well-being. For more information, email chep@rand.org.

Contents

Figures

Tables

Summary

The 1994 Northridge earthquake in Southern California was a severe seismic event. As with previous similar events, the Northridge earthquake prompted a range of policy responses designed to improve the resilience of buildings and other structures in California to future seismic events. The impact of the Northridge earthquake on hospitals was a particular concern. This led to legislation—Senate Bill (SB) 1953—that focused on improving hospital building standards to enhance resilience to seismic events. Since the passage of SB 1953, hospitals have been assessing the implications of the legislation for their business planning and implementing structural and nonstructural upgrades to comply with the standards. These upgrades have been expensive for hospitals. The potential costs of SB 1953 have raised significant concerns regarding both the financial burden the legislation places on hospital systems and the opportunity costs associated with hospitals investing large pools of capital in implementing seismic upgrades.

This report builds on earlier RAND Corporation research to update estimates of the costs to hospitals of future compliance with SB 1953, with a particular focus on the 2030 deadline for ensuring continued operations postevent. In addition to generating direct estimates of the costs of retrofitting or rebuilding noncompliant buildings, the report also assesses the affordability of compliance based on recent hospital financial data. Given the challenges of cost and affordability, this report also characterizes a range of policy alternatives that could be implemented alone or in combination to ease the compliance challenge while building seismic resilience in California hospitals.

Results from quantitative and qualitative analyses indicate that, despite multiple decades of investment in SB 1953 seismic compliance projects, California hospitals still face a significant financial obligation between now and 2030. Depending on assumptions regarding preferred compliance strategy (retrofit versus new construction) and future cost escalation, the statewide price tag for compliance could range from $34 billion to $143 billion. A significant proportion of hospitals are already experiencing some degree of financial stress, and the burden of future compliance is likely to exacerbate this stress.

There are a variety of options for providing regulatory relief or flexibility to hospitals, including providing public subsidies to share the costs of compliance or reduce financing costs, offering additional flexibility in the timing of compliance deadlines, rethinking what it means for hospitals to remain operational postevent, and streamlining the administrative processes associated with seismic compliance projects. Regardless, decisionmaking regarding future changes, if any, in SB 1953 requirements would be aided by more-comprehensive analysis of the benefits of the policy, as well as improved data collection and reporting regarding seismic projects and their costs.

Acknowledgments

We thank Andrew Lauland (RAND), Mireille Jacobson (University of Southern California, Leonard Davis School of Gerontology), and James Langabeer II (University of Texas Health Sciences Center at Houston) for serving as expert peer reviewers of this report and providing comments that improved the analyses and their presentation.

We acknowledge the members of the external project advisory committee, composed of hospital representatives, former hospital executives, and consulting engineers, for their participation and willingness to provide their perspectives on the scope of the study and key issues associated with Senate Bill (SB) 1953 implementation. In addition, we thank the hospital staff who participated in interviews or provided project cost data to RAND in support of various analyses. Similarly, we thank the staff of California's Office of Statewide Health Planning and Development (OSHPD) for multiple discussions of SB 1953 implementation, as well as for facilitating access to OSHPD data.

This study was funded by the California Hospital Association.

Abbreviations

AB	Assembly bill
CAH	Critical Access Hospital
CHA	California Hospital Association
CHNA	community health needs assessment
DOT	California Department of Transportation
DSOD	Department of Water Resources Division Safety of Dams
FEMA	Federal Emergency Management Agency
MHS	multihospital system
NPC	nonstructural performance category
OSHPD	California's Office of Statewide Health Planning and Development
SB	Senate bill
SCU	Seismic Compliance Unit
SPC	structural performance category

1. Introduction

Problem Overview

The Northridge earthquake was among one of the more severe seismic events to affect California during the 20th century. The 6.7-magnitude event occurred on January 17, 1994, with an epicenter located approximately 20 miles northwest of downtown Los Angeles. The earthquake caused widespread damage in Southern California, including 57 deaths, approximately 8,700 injuries, and $20 billion in direct economic damage.[2] A broad range of infrastructure systems were damaged or disrupted by the earthquake, including transportation (e.g., roads and bridges), public buildings (e.g., schools), and hospitals. As with previous events, including the 1989 Loma Prieta earthquake and the 1971 Sylmar earthquake, Northridge prompted a range of policy responses designed to further increase the resilience of California to future seismic events.

Although none of the deaths was directly attributable to hospital failures, structural or otherwise,[3] the impact of the Northridge earthquake on hospitals was a particular concern. A reported 11 hospitals suffered structural or nonstructural damage, such as loss of electrical and water services.[4] In some cases, this damage made the facility unusable, necessitating the eventual replacement of the building. Eight acute care hospitals in Los Angeles County were evacuated (9 percent of the 91 hospitals in the county),[5] either partially or fully, necessitating the transfer of existing patients and diversion of new patients to other facilities.[6] However, despite damage to these facilities, Southern California had sufficient capacity to meet immediate demand through other hospitals.[7]

The damage to hospitals highlighted their continued vulnerabilities to seismic events and the potential health and safety issues that could arise because of damaged hospitals. In particular, the age of hospital buildings, and thus the building codes under which they were built, was identified

---

[2] KPCC, "The Northridge Earthquake, 20 Years Later," Southern California Public Radio, 2004.

[3] D. P. Goldman, M. Vaiana, and J. A. Romley, "The Emerging Importance of Patient Amenities in Hospital Care," *New England Journal of Medicine*, Vol. 363, No. 23, 2010, pp. 2185–2187.

[4] J. Cheevers and A. Abrhamson, "Earthquake: The Long Road Back; Hospitals Strained to the Limit by Injured; Medical Care; Doctors Treat Quake Victims in Parking Lots; Details of Some Disaster-Related Deaths Are Released," *Los Angeles Times*, January 19, 1994.

[5] Acute care hospitals are those that provide 24-hour inpatient care, including multiple basic health care services, such as medical, nursing, surgical, anesthesia, laboratory, radiology, pharmacy, and dietary services. See Appendix A for descriptions of hospital categories used in this study.

[6] C. H. Schultz, K. L. Koenig, and R. J. Lewis, "Implications of Hospital Evacuation After the Northridge, California, Earthquake," *New England Journal of Medicine*, Vol. 348, No. 14, 2003, pp. 1349–1355.

[7] Kathleen J. Tierney, "Social Impacts and Emergency Response," in John F. Hall, ed., *Northridge Earthquake, January 17, 1994: Preliminary Reconnaissance Report*, Oakland, Calif.: Earthquake Engineering Institute, 1994, pp. 87–93.

as a factor contributing to damage in many cases.[8] This led to legislation adopted in June 1994: Senate Bill (SB) 1953,[9] an update to prior legislation that focused on improving hospital building standards to enhance resilience to seismic events. SB 1953 established a 2008 deadline for ensuring the life safety (i.e., noncollapse) of acute care hospitals during seismic events, as well as a 2030 deadline for ensuring continued operations postevent.[10] Since the bill's passage, subsequent policy revisions have been implemented. The original 2008 deadline shifted over time to 2020, yet the 2030 deadline has, to date, held firm. In response, hospitals have implemented both structural and nonstructural upgrades to comply with the standards. These upgrades have been implemented at significant cost to hospitals.

These costs have raised concerns among hospital owners and operations about the financial burden they place on hospital systems and the opportunity costs associated with hospitals investing large pools of capital in implementing seismic upgrades. Capital dedicated to upgrades becomes unavailable for other capital investment or operational improvements that could benefit hospital operations, service provision, and the health and well-being of Californians. On the other hand, in the event of a major seismic event, there is a public benefit to having hospitals that can remain operational and continue to provide health care services. Given that, a central question that persists around SB 1953 is what the total costs of the legislation, both direct and indirect, are, as well as how these costs might vary by hospital type, condition relative to the legislative standards, and other hospital characteristics. Although information about the potential benefits of meeting the requirements of SB 1953 is relevant to evaluating the net impact of the legislation, the assessment of benefits was beyond the scope of this study.

Study Objectives

The objective of this study is to inform decisionmaking around current and future implementation of SB 1953 by analyzing both the current status of compliance across the state and the future costs associated with complying with SB 1953 requirements through the 2030 deadline. These costs are placed in a larger context of hospital planning, which necessarily involves consideration of a broad range of factors. The study was designed to reflect the actual experiences of hospitals to date, while also capturing the future obligations that hospital owners and operators face in the decade ahead. In so doing, the study assumes that efforts to comply

[8] Jean Lin Pao, Michael Siewart, Jeremy Creelan, and Henry Mere, *Preparing for the "Big One": Saving Lives Through Earthquake Mitigation in Los Angeles, California*, Washington, D.C.: U.S. Department of Housing and Urban Development, 1995.

[9] California Senate Bill 1953, An Act to Amend Section 18938 of, and to Add Article 8 (Commencing with Section 15097.100) and Article 9 (Commencing with Section 15097.125) to Chapter 1 of Division 12.5 of, the Health and Safety Code, Relating to Building Standards, and Making an Appropriation Therefor, September 21, 1994.

[10] *Life safety* refers to the life and physical well-being of individuals. It is codified within the *Life Safety Code* (*NFPA 101*), which is a U.S. standard pertaining to hazards to human life in buildings, public and private conveyances and other human occupancies. National Fire Protection Association, *NFPA 101, Life Safety Code*, Quincy, Mass., 2018.

with the SB 1953 requirements for 2020 have effectively been completed or are in the process of being completed. Therefore, planning and investment by hospitals is now oriented toward the 2030 deadline, which is associated with fundamentally different requirements. In addition, the study sought to look beyond hospital owners and operators to integrate qualitative and quantitative information from other stakeholders and analyses.

This study seeks to improve on prior work by pursuing three research pathways:

- First, this study represents an update to two prior studies conducted by the RAND Corporation (published in 2002 and 2007 by the California HealthCare Foundation).[11] This update analyzes the likely magnitude of the costs of both structural and nonstructural improvements for California hospitals over the next 11 years as they implement actions to comply with SB 1953. Hence, the costs of compliance associated with achieving the SB 1953 2020 deadline are treated as sunk costs and are excluded from the analysis. In addition, this study acknowledges the uncertainties associated with the costs of compliance and identifies opportunities for reducing those uncertainties.
- Second, these costs are placed in a broader context of affordability. This includes analysis of the current financial health of California hospitals using an independent metric of financial stress, as well as the analysis of the impacts of SB 1953 compliance on financial health.
- Third, the study compares the requirements and funding arrangements for SB 1953 with other seismic resilience policies targeting other types of critical infrastructure. In addition, the study analyzes an illustrative set of alternative policy options and their potential benefits and trade-offs to seed policy discussions regarding future SB 1953 implementation and potential changes to the legislation.

In building on past work, this study also explicitly recognizes that acute care hospitals represent just one element of society's health care infrastructure, which includes outpatient clinics, emergency medical services, and public health infrastructure, to name a few. Therefore, enhancing the resilience of hospitals and health care delivery to seismic events should be pursued in conjunction with efforts to enhance broader community and infrastructure resilience.

How This Report Is Organized

Chapter 2 presents background information on the history of seismic safety requirements for hospitals, the origins of SB 1953, and subsequent legislation. It also discusses the impact of SB 1953 on hospitals since passage in 1994 and summarizes past work by RAND on the costs of compliance.

Chapter 3 presents the overall analytic approach used in the study, including how the RAND research team engaged hospitals and other stakeholders and the sources of data used in the study.

Chapter 4 discusses the decisionmaking context associated with SB 1953 as hospitals make decisions regarding capital investments over the next decade. This includes identification of the

[11] Charles Meade, Jonathan Kulick, and Richard Hillestad, *Estimating the Compliance Costs for California SB 1953*, Oakland: California HealthCare Foundation, 2002; Charles Meade and Jonathan Kulick, *SB1953 and the Challenge of Hospital Seismic Safety in California*, Oakland: California HealthCare Foundation, 2007.

key factors that hospitals must weigh in determining how and when to undertake seismic compliance projects.

Chapter 5 presents updated cost estimates for compliance with SB 1953. We focus on post-2020 compliance costs, assuming that the costs associated with the disposition of SPC-1 (level 1 of the structural performance category) buildings will have largely been addressed by 2020. Hence, the analysis focuses on the costs of compliance for the remaining stock of SPC-2 buildings and necessary nonstructural performance category (NPC) work to bring buildings into compliance by 2030.

Chapter 6 focuses on the analysis of the affordability of SB 1953. This includes a qualitative discussion of the various factors and trends within the U.S. industry and California, specifically, that influence affordability. In addition, we apply a previously published financial stress index—the Z score—to categorize hospitals based on their financial health. This allows the costs of compliance to be placed in the larger context of affordability. It also enables the identification of types of hospitals that are likely to be disproportionately affected by SB 1953, as well as the additional financial stress that compliance with SB 1953 could generate.

Chapter 7 examines seismic risk management policies for other types of critical infrastructure to place SB 1953 in a broader context. In particular, the chapter analyzes how infrastructure ownership influences access to financing for capital investments in seismic upgrades.

Chapter 8 discusses alternative policy approaches to SB 1953 that span subtle modifications to the existing legislation and requirements to more-substantial reforms. The chapter begins by exploring the structure of seismic policies for other types of critical infrastructure—water resources management, dams, and transportation infrastructure. This reveals that such infrastructure is often publicly owned and operated and thus can draw on public funds as a financial backstop for funding seismic upgrades. The chapter continues by presenting a framework for the analysis of alternatives, based on various assessment criteria. This framework is then applied to a portfolio of policy alternatives to identify benefits and trade-offs that can inform future deliberations regarding hospital seismic safety policy development.

Chapter 9 summarizes the findings and offers additional recommendations to improve future efforts to understand the implications of SB 1953 and the resilience of the California health care system more broadly. In addition, three appendixes provide background information on the types of hospitals used in this study, major seismic events since the Northridge earthquake, and details regarding past changes in seismic safety legislation.

2. Study Background

History of Hospital Seismic Safety Legislation

Origins of the Alfred E. Alquist Act

The modern incarnation of seismic building standards for California hospitals dates back almost 50 years, to the 1971 Sylmar earthquake in the San Fernando Valley of Southern California. The 6.5-magnitude event killed 64 people and caused extensive damage to the Olive View Medical Center and the Veterans Administration Hospital. Power outages led to the failure of life support systems, and several structures collapsed, causing significant loss of life. In particular, buildings erected prior to 1933, when improved seismic construction practices were implemented, were damaged. Three buildings that had been completed in 1970 at Olive View and designed to be earthquake resistant also experienced the collapse of basements, lower stories, and car parks. Olive View was demolished in 1973, and a new facility was constructed in its place, which opened in 1987.

Given the loss of life and damage to property caused by the Sylmar earthquake, the California legislature passed the Alfred E. Alquist Hospital Facilities Seismic Safety Act of 1973.[12] The basic premise of the legislation was described as follows:

> It is the intent of the Legislature that hospital buildings that house patients who have less than the capacity of normally healthy persons to protect themselves, and that must be reasonably capable of providing services to the public after a disaster, shall be designed and constructed to resist, insofar as practical, the forces generated by earthquakes, gravity, and winds. In order to accomplish this purpose, the office shall propose proper building standards for earthquake resistance based upon current knowledge, and provide an independent review of the design and construction of hospital buildings.[13]

The Alquist Act, which was subsequently updated in 1983, required that new construction meet more-stringent building codes. In so doing, the act sought to enhance the resilience of future hospitals in the state and, eventually, as older hospital buildings were retired, raise the overall seismic performance of hospital buildings. The Alquist Act was the first in what would eventually become a long line of legislative and regulatory actions on hospital seismic safety, the most significant of which are highlighted in Table 2.1 (see also Appendix B for a more comprehensive list).

---

[12] California Senate Bill 519, Alfred E. Alquist Hospital Facilities Seismic Safety Act of 1973, 1973.

[13] State of California, California Health and Safety Code, Division 107: Statewide Health Planning and Development, Part 7: Facilities Design Review and Construction, 2019.

Table 2.1. Key Seismic Safety Requirements and Policy Changes (1973–2019)

Year	Policy Change	Implications
1973	Alquist Act passed	Required new acute care hospitals to be designed and constructed to remain standing and operational after a major seismic event. Administration of the new requirements was delegated to OSHPD.
1994	SB 1953	Amended the Alquist Act to establish structural and nonstructural performance categories. Established deadlines of January 1, 2008, for survivability (SPC-2) and January 1, 2030, for continued operations (>SPC-2 and >NPC-3). Mandated that hospitals submit a seismic evaluation to OSHPD for review and approval by January 1, 2001.
2000	SB 1801	Provided an opportunity for a five-year extension on the January 1, 2008, deadline (to January 1, 2013).
2007	SB 306	Provided an opportunity for a seven-year extension on the January 1, 2013, deadline (to January 1, 2020) for city or county hospitals or hospitals that meet strict financial hardship criteria.
2007	Hazus reassessment program	Enabled hospitals determined to be at lower seismic risk because of an application of Hazus to be reclassified to SPC-2, and therefore have until January 1, 2030, to meet other SB 1953 requirements.
2011	SB 90	Provided an opportunity for a seven-year extension on the January 1, 2013, deadline (to January 1, 2020) if necessary planning and reporting requirements were met (including mandatory Hazus assessment).
2016	SPC-4D added to the California Building Standards Code	Enabled nonconforming SPC-1 or SPC-2 buildings to be upgraded to this new category so they can provide acute care services beyond 2030 based on structural requirements (equivalence to the 1980 building code) or seismic performance (seismic evaluation).
2019	AB 2190	Provided an opportunity for extensions (two-year or four-year extensions, depending on compliance strategy) on the January 1, 2020, deadline if certain conditions were met.
2019	NPC-4D added to the California Building Standards Code	Allowed hospitals to be in compliance with the NPC January 1, 2030, deadline provided that they, at a minimum (level 1), meet the NPC-3 requirements and develop an operational plan for repair of nonstructural damage sustained during a seismic event. Hospitals are still are required to meet the NPC-5 requirement.

SOURCE: See OSHPD, "Changes to Seismic Safety Laws," webpage, undated.
NOTES: OSHPD = California's Office of Statewide Health Planning and Development; AB = Assembly bill. For the California Building Standards Code, see California Building Standards Commission, "California Building Standards Code," webpage, undated.

The Northridge Earthquake and Passage of SB 1953

In 1994, the Northridge earthquake led to the reevaluation of the Alquist Act. The damage experienced by hospitals and the associated loss of life in hospital facilities, though limited, led policymakers to conclude that additional steps were needed to enhance the seismic resilience of California hospitals. A key observation was that hospitals built after the 1973 Alquist Act generally performed well structurally, although they experienced nonstructural damage. Legislative action culminated in SB 1953. This amendment to the Alquist Act included provisions for not only further improving the standard for new construction but also mandating structural upgrades to the *existing* stock of acute care buildings in the state. The legislation determined that, "[a]fter January 1, 2008, any general acute care hospital building that is determined to be a potential risk of collapse or pose significant loss of life shall only be used for nonacute care hospital purposes." Hospitals unable to meet the deadline could apply for a five-year extension, to 2013, by explaining why they were unable to meet the deadline.

The requirement that existing hospitals also meet updated seismic performance standards represented a significant departure from prior legislation. All hospitals in the state were given a classification based on seismic risk—the SPC and the NPC (see Table 2.2). The highest-risk buildings were classified as SPC-1, meaning there was an unacceptable risk of collapse during a seismic event posing life safety risks. Hospitals were left with two options: Retrofit existing general acute care buildings or take those buildings out of acute care service. Moreover, deadlines were established for addressing noncompliant buildings. SPC-1 buildings were originally mandated to be upgraded to SPC-2 by 2008 or taken out of service. Meanwhile, all other buildings were required to be compliant to SB 1953 requirements by 2030. By 2001, 1,027 hospital buildings were classified as SPC-1, which was approximately 40 percent of all general acute care buildings in the state.[14]

Table 2.2. SPC Definitions

SPC Level	Description	Deadlines
SPC-1	Buildings posing significant risk of collapse and danger to the public	May not be used beyond January 1, 2020
SPC-2	Buildings in compliance with the pre-1973 California Building Standards Code or other applicable standards, but not in compliance with the structural provisions of the Alquist Act	May not be used beyond January 1, 2030
SPC-3	Buildings in compliance with the structural provisions of the Alquist Act, utilizing steel moment-resisting frames in regions of high seismicity as defined in Section 4.2.10 of the California Building Standards Code and constructed under a permit issued prior to October 25, 1994	May be used until January 1, 2030
SPC-4	Buildings in compliance with the structural provisions of the Alquist Act but that might experience structural damage, which might inhibit ability to provide services to the public following strong ground motion	May be used beyond January 1, 2030
SPC-5	Buildings in compliance with the structural provisions of the Alquist Hospital Facilities Seismic Safety Act and reasonably capable of providing services to the public following strong ground motion	May be used beyond January 1, 2030

SOURCES: OSHPD, "Seismic Performance Ratings," webpage, undated. For the California Building Standards Code, see California Building Standards Commission, "California Building Standards Code," webpage, undated.

When initially passed, the legislation was designed to increase the resilience of California hospitals—ensuring life safety over the near term (i.e., by 2008) and resilient operations over the long term (i.e., post-2030). Moreover, with the latter deadline set for 2030, by that point, many of the aging, noncompliant buildings would be at least a half century old and thus likely reaching the end of their life spans. However, as it happened, SB 1953 created incentives for hospitals to

[14] D. Alesch, W. Petak, and L. Arendt, *Natural Hazard Mitigation Policy: Implementation, Organizational Choice and Contextual Dynamics*, New York: Springer, 2012.

assess their exposure to the legislation, decide on appropriate courses of action, and implement those actions. By January 1, 2001, hospitals had completed the seismic evaluations mandated by SB 1953 and approved by California's OSHPD, which enabled the SPC and NPC classifications of buildings. This revealed the scale of the effort and investment that would be required to achieve compliance, particularly with respect to SPC-1 buildings. The fact that SB 1953 was an unfunded mandate placed additional pressure on hospitals, forcing them to reconcile the implications of the legislation with their capital investment strategies, future service provision, and financial health.[15]

By the early 2000s, the challenge of completing such a massive overhaul of hospital infrastructure over such a short time became increasingly apparent. In 2000, SB 1801 authorized hospitals to receive a five-year extension to the 2008 deadline, provided certain conditions were met.[16] This effectively pushed the deadline for addressing SPC-1 buildings to 2013. Although extensions did provide additional time for hospitals to plan and implement upgrades, and might have reduced, but not eliminated, the cost burden, they did not result in full compliance. By 2009, one year after the original deadline, there were still 825 SPC-1 buildings in the state.[17]

A more substantive policy shift came in the form of the Hazus Reassessment Program, which was authorized by OSHPD in November 2007. The Hazus program allowed hospitals with collapse-hazard buildings to voluntarily use a state-of-the-art technology called Hazus (Multi-Hazard Loss Estimation Methodology, Earthquake Module) to reevaluate their seismic risk.[18] This led OSHPD to pursue a more refined classification of hospital buildings, reducing the number of SPC-1 buildings by half.[19] In 2010, OSHPD implemented an emergency regulation that raised the threshold for SPC-1 classification, which allowed an even larger number of buildings to be reclassified to SPC-2. By 2010, Hazus and extensions left just 145 SPC-1 buildings in need of compliance actions in advance of the 2013 deadline.[20]

In parallel with the Hazus program, deadlines for SB 1953 compliance also continued to shift. In 2007, SB 306 provided an additional seven-year extension (to January 1, 2020) for city and county hospitals if they were not able to afford the costs of achieving the 2013 deadline.[21] This was followed in 2011 by SB 90, which broadened the eligibility requirements for the seven-year extension, making 2020 the effective deadline for addressing noncompliant SPC-1

---

[15] M. J. McCue and J. M. Thompson, "Hospital Compliance with a State Unfunded Mandate: The Case of California's Earthquake Safety Law," *Hospital Topics*, Vol. 90, No. 4, 2012, pp. 91–97.

[16] California Senate Bill 1801, An Act to Amend Section 130060 of the Health and Safety Code, Relating to Health Facilities, September 29, 2000.

[17] Alesch, Petak, and Arendt, 2012.

[18] Tom Chang and Mireille Jacobson, "Natural Disaster Management: Earthquake Risk and Hospitals' Provision of Essential Services in California," in Daniel P. Kessler, ed., *Regulation Versus Litigation: Perspectives from Economics and Law*, Chicago: University of Chicago Press, 2010, pp. 301–320.

[19] C. Poland, "California's Wise Retreat from Seismic Overreaction," *Healthcare Design*, May 31, 2008.

[20] Alesch, Petak, and Arendt, 2012.

[21] California Senate Bill 306, An Act to Amend Section 129765 of, and to Add Section 130061.5 to, the Health and Safety Code, Relating to Health Facilities, October 13, 2007.

buildings.[22] More recently, AB 2190 allowed for further extensions (to 2022 or 2024) for those hospitals that are in the process of complying with the 2020 deadline but are unlikely to complete necessary projects by then.[23] These various actions created opportunities for some flexibility in the timing of seismic upgrades by hospitals, provided that they were diligent in completing the administrative criteria for extension eligibility.

Although the 2020 deadline applies to life safety, the 2030 deadline places additional requirements on hospitals to enable continued operations during and after a seismic event. Buildings that are categorized as SPC-3, SPC-4, or SPC-5 are all compliant and represent buildings that were built to a particular code deemed adequate to maintain structural performance. However, SPC-2 buildings, which have a low risk of collapse, must nevertheless perform seismic upgrades. SPC-2 buildings cannot be upgraded to SPC-3 or SPC-4; therefore, as originally intended, SPC-2 buildings must be retrofitted to SPC-5. Hence, much of the focus of hospitals post-2020 will be on meeting this next level of compliance associated with the SPC-2 buildings within hospital facilities—many of which were constructed in the 1950s, 1960s, and 1970s and pose nontrivial challenges to structural upgrades.

Concerns in recent years about the equity implications of upgrading SPC-2 buildings to SPC-5 led to the reevaluation of the original policy. Specifically, by requiring buildings to meet SPC-5, the law would require retrofitting resulting in higher performance than other buildings that are considered compliant. This observation led to the introduction of a "damage control" classification, designated SPC-4D within the 2016 California Building Standards Code, which took effect as of January 1, 2017.[24] SPC-4D enables buildings built prior to 1973 to be retrofitted to CBC (California Building Code) 1980 or ASCE (American Society of Civil Engineers) 41 standards rather than meeting the SPC-5 category. This has the potential to reduce the costs associated with seismic upgrades but also necessitates additional project evaluation and assessment work to determine whether a building can be retrofitted to SPC-4D and how.

SB 1953 also created requirements for nonstructural seismic upgrades (see Table 2.3). The NPC-1 and NPC-2 categories are consistent with ensuring the life safety of buildings by increasing the resilience of those utilities necessary for evacuation of a hospital facility (e.g., emergency lighting to enable egress). Therefore, these requirements had to be met quite early (by 2002). However, the higher categories are more focused on enabling hospitals to continue to provide acute, and critical, care services after an event by bracing and anchoring of equipment, utilities, and services. For example, NPC-3 requires enhanced resilience of laboratory areas,

---

[22] California Senate Bill 90, An Act to Amend Section 130060 of the Health and Safety Code, and to Amend Sections 14105.281, 14166.115, and 14167.10 of, to Amend and Repeal Section 14166.245 of, and to Add and Repeal Article 5.226 (Commencing with Section 14168.1) and Article 5.227 (Commencing with Section 14168.31) of Chapter 7 of Part 3 of Division 9 of, the Welfare and Institutions Code, Relating to Health, Making an Appropriation Therefor, and Declaring the Urgency Thereof, to Take Effect Immediately, April 13, 2011.

[23] California Assembly Bill 2190, An Act to Add Sections 130062 and 130066 to the Health and Safety Code, Relating to Hospitals, September 22, 2018.

[24] California Code of Regulations, Title 24, published July 1, 2016, effective January 1, 2017; see California Building Standards Commission, "California Building Standards Code," webpage, undated.

pharmaceutical services, radiological services, and central and sterile supply areas. Meanwhile, NPC-4 focuses on architectural, mechanical, and electrical equipment. To reduce vulnerability to disruptions of external infrastructure, NPC-5 requires on-site storage of sufficient water, wastewater, and fuels for a hospital to be self-sufficient for a few days.

Although NPC projects are generally undertaken at lower cost than SPC projects, they do pose challenges. A larger number of buildings and a larger total area of hospitals are affected by NPC requirements, and completing NPC work can cause disruption to normal operations. Hence, once the 2008 SPC deadline shifted to 2013, and then 2020, it made sense to shift the NPC-4 and NPC-5 deadlines as well. That allowed greater coordination between SPC and NPC projects and thus reduced the costs of compliance.

Table 2.3. NPC Definitions

NPC Level	Description	Deadline
NPC-1	These are buildings with equipment and systems that do not meet the bracing and anchorage requirements of any other NPC.	—
NPC-2	The following systems are braced or anchored in accordance with Part 2, Title 24 of the California Building Standards Code: communications systems, emergency power supply, bulk medical gas systems, fire alarm systems, and emergency lighting equipment and signs in the means of egress.	January 1, 2002
NPC-3	The building meets the criteria for NPC-2. Also, in critical care areas, clinical laboratory service spaces, pharmaceutical service spaces, radiological service spaces, and central and sterile supply areas, the following components meet the bracing and anchorage requirements of Part 2, Title 24: nonstructural components (in Part 2, Title 24), equipment and fire sprinkler systems.	January 1, 2008
NPC-4	The building meets the criteria for NPC-3. Also, all architectural, mechanical, and electrical systems, components, and equipment and hospital equipment meet the bracing and anchorage requirements of Part 2, Title 24.	January 1, 2020, or January 1, 2030, depending on SPC rating
NPC-5	The building meets the criteria for NPC-4. Also, on-site supplies of water and holding tanks for sewage and liquid waste, sufficient to support 72 hours of emergency operations, are integrated into the building plumbing systems in accordance with the California Plumbing Code.	January 1, 2030

SOURCES: OSHPD, "Seismic Performance Ratings," webpage, undated. For the California Plumbing Code, see California Building Standards Commission, *2016 California Plumbing Code: Entire Code with Errata Pages*, California Code of Regulations, Title 24, Part 5, Ontario, Calif.: International Association of Plumbing and Mechanical Officials, July 2016.
NOTE: — = not applicable.

Nevertheless, by 2016, hospitals were seeking additional flexibility with NPC requirements. Just as SPC-4D provided a mechanism for greater flexibility in bringing SPC-1 and SPC-2 buildings into compliance, OSHPD's most recent policy modification is the introduction of a new NPC category—NPC-4D (see Table 2.1).[25] This new standard, which is a feature of the

[25] Paul A. Coleman, *2019 California Building Standards Code Update for the California Hospital Association*, Sacramento, Calif.: Office of Statewide Health Planning and Development, 2018.

2019 California Building Standards Code, effectively represents an extension of the NPC-3 category. Hospitals can satisfy the NPC-4D requirement by, at a minimum (or level 1), meeting the NPC-3 requirements and submitting a management plan with contingencies to address residual disruption to services in the event of a seismic event. At level 2, utilities to level 1 areas are anchored and braced to provide service to patient, surgical, obstetrical, and ground floors. Level 3 includes level 2, plus anchoring and bracing of additional systems designated by the hospital as being necessary to maintain operations. Ultimately, NPC-4D provides more flexibility in terms of nonstructural compliance, provided that an adequate plan exists to manage damage and disruptions that arise within a building because it is not NPC-5.

Summary of Past Findings

The Alquist Act, in general, and SB 1953, specifically, have been features of California's regulatory landscape for 45 years and 25 years, respectively. Despite the bill's long history and the significant impact it has had on strategic planning and capital investment among the state's hospitals, there has been relatively little analysis of the policy and its consequences—either before its implementation or routine monitoring of its impact. Although clearly developed as a mechanism for protecting the health and safety of Californians in the event of a significant earthquake, no formal cost-benefit analysis of SB 1953 was conducted prior to its implementation. Therefore, understanding of the costs, opportunities, and challenges has largely emerged through learning by doing as hospitals have pursued seismic compliance projects.

Assessing the benefits of the policy directly is difficult because of the paucity of large-scale seismic events that have occurred since 1994 (see Appendix C), as well as the fact that implementation of seismic upgrades remains on ongoing process. The events that have occurred since Northridge have had little adverse impact, although this is due to the modest magnitude of the events or their remote locations, away from dense population centers. The benefits of the standards therefore remain largely untested against actual seismic events.

Nevertheless, the risk of future events that exceed structural or nonstructural limits of California's hospitals remains. For example, the 7.0 earthquake that struck Anchorage, Alaska, on November 30, 2018, caused significant damage to two hospitals, although emergency departments at both remained open.[26] Although not directly transferable to the California context, this recent event demonstrates the potential vulnerabilities that exist in hospitals and, therefore, the benefits of ensuring that hospital structural and nonstructural elements and operations are robust to seismic risk.

---

[26] Anne Hillman, "Emergency Rooms Open, Non-Critical Appointments Canceled at Anchorage Area Hospitals," Alaska Public Media, November 30, 2018.

Only a small number of studies estimate the costs of SB 1953. The California Hospital Association (CHA) completed a report in 1999,[27] and RAND conducted two studies in the subsequent few years. We briefly revisit these prior efforts to place the current study in context and to evaluate what has changed in the time since they were completed.

RAND previously undertook two studies of the costs of compliance for SB 1953, published in 2002 and 2007 by the California HealthCare Foundation.[28] Both studies narrowly address the direct costs of compliance and assess the costs associated with SPC-1 buildings. This focus is a function of two factors. First, the studies prioritized the (then) near-term deadlines of 2008 and 2013. Second, the prevailing assumption at the time was that hospitals would address noncompliant SPC-1 buildings by reconstructing them. That would have left only a modest number of SPC-2 buildings (representing 8.3 percent of the statewide inventory in 2006) to address by 2030. For example, the 2002 report identifies five times as many SPC-1 buildings as SPC-2.[29]

However, implementation of the Hazus program soon followed the 2007 report, which led to the reclassification of the majority of SPC-1 buildings by OSHPD. In so doing, the Hazus program significantly reduced the compliance burden for SPC-1 buildings but significantly increased the burden of SPC-2 buildings and effectively shifted the bulk of the compliance challenge to the 2030 deadline. Therefore, the 2002 and 2007 RAND studies represent a fundamentally different era in terms of seismic compliance when compared with the present. This makes comparison of the total cost estimates across the entire time period challenging.

2002 RAND Report

The original RAND report was published in 2002,[30] eight years after the passage of SB 1953. By that point, it was apparent that the legislation as written would affect approximately 50 percent of the state's hospitals. The required scale of construction would be unprecedented in the history of California. The following are some of the highlights from the 2002 report:

- Total costs of compliance were estimated at $41.7 billion, with $41.0 billion for retrofitting, rebuilding, or replacing SPC-1 and SPC-2 buildings and $0.7 billion for NPC work. This was almost twice the $23.8 billion estimated by the CHA in 1999. However, if one excluded construction that was not strictly seismic-related, the cost dropped to $3 billion.
- It was assumed that most of the costs of compliance would be addressed through normal recapitalization of hospital assets, given natural turnover of building stock on a 40- to 50-year period.
- The aging of California's hospitals, combined with their slow rate of recapitalization, was argued to hinder the ability of hospitals to keep pace with evolving trends in health care

---

27 J. Gillengerten, "Estimated Cost of Structural Retrofits," paper presented at the SB 1953 Seismic Compliance Meeting, California Healthcare Association, Sacramento, March 30, 1999.

28 Meade, Kulick, and Hillestad, 2002; Meade and Kulick, 2007.

29 Meade, Kulick, and Hillestad, 2002.

30 Meade, Kulick, and Hillestad, 2002.

practice and service delivery, particularly the shift toward greater use of outpatient services.

2007 RAND Report

Several years after its original report, RAND updated its analysis of the costs of compliance.[31] During that time, there were few changes in the numbers of noncompliant buildings, suggesting that efforts to address seismic deficiencies were progressing slowly. The following are some highlights from the 2007 report and some of its differences with the prior analysis:

- The report estimated that the total costs of compliance at $45 billion–$110 billion but noted that the financing of construction projects could effectively double these costs. The report also shifted away from separating seismic and nonseismic construction costs (as in the 2002 report), because there was no strategy available to hospitals that would allow them to undertake only seismic-specific projects.
- Construction of new replacement buildings was assumed to be the dominant compliance strategy because it was less disruptive and created opportunities for modernization.
- Costs of construction were estimated to be 40 percent higher than in other states, owing to California having the most-stringent seismic regulations of any state and competition for the specialist contractor and labor requirements for hospital and seismic construction.
- It was estimated that only half of noncompliant hospitals would meet the 2020 deadline and that many would not comply with the 2030 deadline. Instead, an estimated 30 years would be needed to achieve full compliance, which implied a final date of 2040.
- Of the 28 hospitals that closed in California between 2001 and 2005, 19 had SPC-1 buildings, although the report did not directly attribute closures to the presence of noncompliant SPC-1 buildings.
- The aging of hospitals, combined with their slow rate of recapitalization, was argued to hinder the ability of hospitals to keep pace with trends in health care practice, particularly the shift toward greater use of outpatient services.
- The 2007 report also identified some policy alternatives, including continuing with business-as-usual implementation of the legislation, modifying the requirements for compliance, or providing funding assistance to hospitals to help them overcome challenges associated with affordability.

Comparing Prior Compliance Cost Estimates

To compare the cost estimates from various prior studies (CHA and the two RAND studies), we applied an inflation factor of 2.3 percent to all estimates to escalate them to 2019 dollars (see Table 2.4).[32] Although the actual costs remain uncertain, this comparison suggests that between 1999 and 2006, the anticipated total costs of SB 1953 increased significantly. After normalizing to constant dollars, medium cost estimates in the 2007 report were almost triple what was

---

[31] Meade and Kulick, 2007.

[32] The 2.3 percent figure represents the average annual increase in the quarterly California consumer price index between 2005 and 2017; see California Department of Industrial Relations, "California Consumer Price Index," webpage, undated.

estimated by the CHA in 1999. The share of this increase that was due to improved understanding of the costs versus differences in methods among studies is unclear.

Table 2.4. Estimates of Total Compliance Costs for SB 1953 from Past Studies, Normalized to Constant (2019) Dollars Using Historical Values of the Consumer Price Index

Report	Cost (billions, current dollars)	Cost (billions, constant 2019 dollars)
CHA (Gillengerten, 1999)	23.8	37.5
RAND (Meade, Kulick, and Hillestad, 2002)	41.7	61.4
RAND (Meade and Kulick, 2007)	45–110	59.1–144.5
RAND (Meade and Kulick, 2007), with finance costs	90–220	118.2–289.0

3. Approach to Analysis

This report extends prior efforts by RAND to assess the costs of compliance with SB 1953 by updating prior analyses to reflect the evolution of seismic efforts in California and offering alternative policy approaches to seismic resilience. In so doing, the report focuses primarily on the future compliance costs California hospitals face as they plan for the 2030 deadline. This chapter summarizes how data and information needed to undertake the study were collected, as well as the sources of those data. Information on specific methods used to undertake different analyses are presented in subsequent chapters in conjunction with the relevant material.

Project Advisory Committee

An external advisory committee was established at the outset of the project. The committee was composed of representatives from ten different hospitals around the state, former hospital executives, and private-sector structural engineers. The committee convened via conference call every other week and served to provide high-level input on key compliance-related issues of concern to hospitals, provide technical clarification on specific seismic requirements and their implications, and identify other stakeholders with relevant technical knowledge who could be engaged in the project.

Engagement with Hospitals and Builders

The research team met with representatives of 13 hospitals in separate meetings between October 2018 and January 2019 to understand hospital decisionmaking related to capital reinvestment and plans, challenges, and costs associated with complying with SB 1953. The interviewees included representatives from for-profit and not-for profit hospitals, ranging from large health care systems composed of multiple hospitals in California or across the country to single hospitals, such as county, district, or Critical Access Hospitals (CAHs). Representatives generally were facilities and construction managers and engineers, and sometimes also were communications, finance, and executive staff. Meetings followed a semistructured discussion protocol that covered the following:

- characteristics of the hospital and the community it serves
- factors that influence decisions about investing in building infrastructure
- actions already taken in response to SB 1953
- plans for complying with future SB 1953 requirements
- cost estimates for complying with SB 1953, broken out by construction cost elements for each of SPC, NPC, or replacement costs
- challenges posed by SB 1953 requirements
- feedback on potential alternatives to SB 1953.

The team also met with representatives from three construction firms to get more-detailed insights into retrofit and replacement projects, including construction approaches and trends, costs (particularly for completed projects), and health care construction market trends.

Engagement with OSHPD

The research team also met with representatives from OSHPD on multiple occasions to obtain data, clarify SB 1953 requirements, and understand the process through which hospitals interact with OSHPD with regard to SB 1953 compliance projects. Data consisted of publicly available information on the compliance status of hospital buildings and on hospital construction projects, and we used custom data on historical SPC and NPC ratings and bed counts in SPC-2 buildings.

Overview of Analytic Tasks

This study was composed of three analytic components, each designed to address an associated research question. Each of these components draws on a variety of data inputs from multiple sources.

Cost Estimation

We focus first on estimating hospitals' future costs of compliance with SB 1953 (see Chapter 5). Although the cost to hospitals does not necessarily reflect the complete social costs of SB 1953, assessing those social costs was beyond the scope of this analysis. Given that most hospital revenue in California comes from Medicare, Medi-Cal, and private health insurers using established reimbursement schemes, providers cannot always easily increase revenue when their costs increase. Estimating the distribution of costs among all stakeholders would require a model of the entire health care financial system.

We generated estimates through the analysis of data from two sources. First, we used data that are publicly available through OSHPD, including data on hospital building characteristics (such as seismic performance classifications maintained by OSHPD) and data on past construction projects (seismic and nonseismic) reported to OSHPD and their associated costs. Second, we collected data directly from hospitals through interviews. Because interviews were possible only for a small subset of California hospitals, subsequent requests for seismic project cost data were also circulated to the CHA membership using a standard template and the completion of data sharing or confidentiality agreements, when necessary.

We also collected data on construction costs and trends from contractors in the industry, consisting of interviews with contractors and estimates published by construction firms in the public domain. These various data sources were used to construct cost models that enabled cost estimation for California hospital buildings, as well as the extrapolation of cost estimates to California as a whole. Consistency checks were subsequently performed, where possible, using independent data and assumptions to confirm that results generated by hospital self-reported data were consistent with industry norms. In addition, we identified other cost drivers of seismic compliance that were difficult to quantify in dollar values but are nevertheless likely to have cost implications for specific hospitals (see Chapter 4).

Affordability Analysis

Next, we analyzed the affordability of compliance with SB 1953 (Chapter 6). Using current financial data reported to OSHPD by hospitals, a financial stress index was used to categorize different hospitals and assess the magnitude of the financial burden that SB 1953 creates for hospitals. This enabled the analysis of how alternative patterns of hospital ownership, trends in reimbursement rates, and access to financial capital affect the affordability of SB 1953 compliance. In addition, estimates of the costs of compliance for California hospitals were integrated into the financial stress index to estimate the additional financial stress hospitals would experience if obligated to achieve compliance. Additional context on affordability was gained through interviews with hospital executives.

Analysis of Policy Alternatives

Finally, we explored what alternative policy approaches could address common barriers to compliance, enhance flexibility in achieving compliance, and reduce overall costs of compliance. In so doing, we first examined approaches to seismic resilience for other types of critical infrastructure (Chapter 7), followed by a more structured analysis of a set of potential policies specific for hospitals (Chapter 8). In addition, hospital representatives were asked during interviews for potential policy alternatives based on their experiences with seismic compliance.

4. SB 1953: Making Decisions for the Next Decade

For a number of years, it has been widely recognized that California's Alquist Act (updated in SB 1953) will transform the state's health care infrastructure because of the demands it places on hospital owners and operators to either retrofit buildings or construct new ones at significant cost. The original assumption that compliance could be achieved over time through the natural depreciation and turnover of buildings proved to be erroneous. Instead, as discussed in Chapter 2, there have been numerous public policy debates over the past two decades, as well as a number of policy changes, in an attempt to accommodate hospital concerns. Yet one critical question remains unresolved: Is SB 1953, and its specific focus on hospitals, the most appropriate policy design for building the resilience of California's health care delivery infrastructure to seismic events?

Consensus on the answer to this question remains elusive among those who own and operate hospitals, regulate hospitals, work in hospitals, or use hospitals. Hospital representatives interviewed as part of this study, for example, expressed concerns regarding the potential effects of compliance with SB 1953 on hospital finances, operations, and capacity. Some explicitly stated their belief that the capital investments being made in SB 1953 compliance could potentially be invested in other ways that could achieve a greater net benefit for Californians. In contrast, other stakeholders have historically opposed efforts over the past two decades to grant hospitals greater flexibility in the timing of seismic upgrades. For example, the California Nurses Association opposed both SB 1661 in 2006 and SB 499 in 2009. The association argued that hospitals have had advanced notice since the original Alquist Act of 1973 that seismic upgrades were needed; thus, additional delays in SB 1953 implementation and compliance were not warranted, given the associated risk to patients and health care workers.[33] Despite these differences of opinion, there is common interest among different stakeholders to provide for the health and well-being of Californians.

Three factors lie at the foundation of this lack of consensus. The first is the manner in which the costs and benefits of the legislation are distributed. Hospitals are charged with shouldering the entire costs of SB 1953 implementation, and hospitals have to pay those costs up front. Benefits are realized only if there is a significant seismic event, and it is not clear whether the potential future benefits of being able to continue to operate without interruption after an event offsets the real, present costs of seismic upgrades or new construction. Meanwhile, the general public would benefit from seismic-resilient hospitals were an event to occur, but the costs to the public are hidden, because they do not pay for seismic improvements directly. Rather, they pay in the form of opportunity costs associated with hospitals investing in seismic resilience when,

---

[33] California Senate Rules Committee, bill analysis and third reading of Senate Bill 1661, Sacramento, April 17, 2006; California Senate Health Committee, bill analysis of Senate Bill SB 499, Sacramento, April 15, 2009.

potentially, they could have invested in other forms of health service delivery that ultimately benefit patients.

The second factor standing in the way of consensus is uncertainty. Although California is at high risk of major seismic events in the future, the exact timing, magnitude, and location are unknown. It is possible to estimate the costs of SB 1953 implementation, albeit imperfectly, but there are fundamental challenges to estimating the benefits of policies designed to protect against events of unknown probability. No formal cost-benefit analysis was conducted when SB 1953 was originally drafted, and none has emerged in the intervening years. This means that hospitals in California have been implementing the requirements of SB 1953 with little prior information on the ultimate implications.

What has been observed, however, is that the challenges encountered in implementing SB 1953 have led to a variety of changes in the requirements and deadlines—often through legislative interventions. Hence, there is now a well-established precedent for changes to SB 1953 when existing requirements are argued to be unreasonable or infeasible. This creates uncertainty regarding whether existing policy requirements should be taken at face value or whether hospitals should assume the requirements will again be subject to change in the future. If the latter, then delaying implementation until there is greater clarity on future policy directions is a rational response.

The third factor is the changing model of health care delivery and the role of hospitals in that model. As noted in the prior RAND studies,[34] the past several decades have witnessed a long-term trend toward greater use of outpatient facilities and services. Nevertheless, hospitals remain an important element of health care delivery, and the types of specialized services they provide cannot be readily replicated. Some hospital representatives argued that the focus of SB 1953 on general acute care hospitals does not fully account for this trend. Rather, this focus creates path dependence around inpatient infrastructure and reduces opportunities to make investments in other delivery models. Therefore, the optimal allocation of capital investment between inpatient and outpatient facilities is unclear.

To further investigate the challenges that hospitals face moving forward, this chapter first highlights the transition in the decisionmaking context for hospitals as they orient toward the 2030 deadline. This is followed by discussion of the current status of compliance for California hospitals vis-à-vis SB 1953 requirements. The chapter concludes by identifying some of the specific factors that influence decisionmaking around compliance strategies.

The Transition from 2020 to 2030

As of December 2018, only 6 percent of California's general acute care buildings still had an SPC-1 designation (Table 4.1), and plans for bringing these buildings into compliance were well underway. Plans consist of a combination of retrofits, new construction, and taking buildings out

[34] Meade, Kulick, and Hillestad, 2002; Meade and Kulick, 2007.

of general acute care service. For a number of hospitals, these projects will exceed the 2020 deadline, with extensions granted from OSHPD. Nevertheless, through a combination of regulatory relief, SPC reclassifications approved by OSHPD, deadline extensions, large capital investments, and retirement of assets, California hospitals will likely make a successful transition to a point where life safety is provided within hospital settings.

As hospitals complete work on their compliance with the 2020 deadline, efforts are now transitioning toward addressing the 2030 deadline for both SPC and NPC compliance. This represents a potentially larger challenge for hospitals, for the following reasons:

- **The number of SPC-2 buildings:** At 631, noncompliant SPC-2 buildings outnumber noncompliant SPC-1 buildings by more than a factor of three. However, given that the original number of SPC-1 buildings was more than 1,000, 631 is modest by comparison.
- **SPC requirements:** Ensuring life safety is a fundamentally different performance objective than maintaining operations and services in the wake of a severe seismic event. The more ambitious objective of continued operations necessitates greater investment in the structural integrity of buildings and therefore creates additional costs above and beyond what is required for an SPC-2 level of compliance.
- **NPC requirements:** The NPC requirements necessitate actions in hospitals that can be quite intrusive, as they require accessing, bracing, and anchoring a diverse array of utilities and services distributed throughout hospital buildings. Moreover, a larger proportion of California hospital buildings are noncompliant with respect to NPC requirements than with SPC requirements.

Given the above, hospital representatives frequently noted that the dominant benefit generated by seismic regulations is likely the prevention of hospital collapse and the protection it affords to the immediate health and safety of hospital workers and patients. With this milestone nearly achieved, the additional benefit achieved by enabling hospitals to continue to operate after an event may be small by comparison. If so, then the return on investment might not be sufficient to justify the large investment still needed for hospitals to comply with the 2030 deadline. This report does not estimate the benefits of SB 1953 and thus does not comment on the merit of such an argument. Nevertheless, despite the 2030 deadline having been established in 1994, some hospitals appear to be framing post-2020 compliance as a new decisionmaking process.

California's Hospital Infrastructure and SB 1953 Compliance Status

California currently has more than 81,000 beds in 418 general acute care hospitals.[35] The size of these hospitals varies substantially, from small, single-building facilities to large, multibuilding campuses. The number of beds per hospital ranges from 10 to more than 800, with an average of about 200.

To be compliant with SB 1953, all buildings on a general acute hospital campus must be SPC-3, SPC-4, SPC-4D, or SPC-5 and NPC-4 (or NPC-4D) and NPC-5. The SB 1953 compliance status of these 418 acute care hospitals as of September 2018 is shown in Table 4.1.

---

[35] OSHPD, "Hospital Building Data," data set, accessed September 2018.

Only 23 of these hospitals (6 percent) have no SPC-1, SPC-2 or NPC-1–NPC-4 buildings and are therefore compliant. Of the 395 hospitals yet to reach compliance, 282 require SPC upgrades, and all 395 require NPC upgrades.

Table 4.1. SB 1953 Compliance Status, by Hospital

Compliance Status	Number of Hospitals
Compliant	23
Noncompliant	395
Has SPC-1 buildings	76
Has SPC-2 buildings	250
Has SPC-1 or SPC-2 buildings	282
Has NPC-1 buildings	29
Has NPC-2 buildings	301
Has NPC-3 or NPC-4 buildings	65

SOURCE: OSHPD, "Hospital Building Data," data set, accessed September 2018.

Compliance statistics are broken out by hospital type and beds in Table 4.2. This table shows that upgrade requirements affect the vast majority of hospital types. However, they disproportionally affect CAHs (100 percent), private hospitals that are not part of multihospital systems (MHSs), public health care district hospitals (100 percent), academic hospitals (100 percent), and hospitals with high Medi-Cal patient-days (98 percent). This means that populations served by these hospitals are more likely to be affected by the 2030 requirements.[36]

The compliance status by building as of September 2018 is shown in Table 4.3. Two-thirds of California hospital buildings require some seismic upgrade, with 815 (26 percent) requiring SPC upgrades and 1,951 (63 percent) requiring NPC upgrades.

[36] See Appendix A for definitions of these categories of hospitals.

Table 4.2. Number and Proportion of Hospitals Affected by SB 1953's 2030 Requirements, by Hospital Type

	All		Have at Least One SPC-1, SPC-2, NPC-1, NPC-2, NPC-3, or NPC-4 Building
Hospital Type	**Hospitals**	**Beds**	**% of Hospitals**
All hospitals	418	81,414	95
CAHs	34	1,467	100
Private—MHS	225	50,280	95
Private—non-MHS	72	14,851	100
Public—non–health care district	19	6,993	95
Public—health care district	38	5,145	100
Academic	11	4,372	100
High Medi-Cal	170	36,971	98

SOURCE: OSHPD, "Hospital Building Data," data set, accessed September 2018.

NOTES: These categories are not mutually exclusive. The private, public, and academic categories are mutually exclusive with each other. The CAH and high-Medi-Cal hospitals are not mutually exclusive with each other, nor with the private, public, and academic categories. For example, a high-Medi-Cal hospital could be a public CAH hospital also, and a private hospital could also be a high-Medi-Cal hospital.

Table 4.3. SB 1953 Compliance Status, by Hospital Building (as of September 2018)

Compliance Status	Number of Buildings
Compliant	339
Noncompliant	2,717
SPC-1	184
SPC-2	631
SPC >2	2,310
SPC unknown	18
NPC-1	141
NPC-2	1,810
NPC-3 or NPC-4	763
NPC unknown	24

SOURCE: OSHPD, "Hospital Building Data," data set, accessed September 2018.

NOTES: Compliant buildings include those classified as SPC-3, SPC-4, and SPC-5 that are also classified as NPC-5. SPC >2 = SPC-3 to SPC-5.

Factors Influencing Hospital Decisionmaking

Seismic compliance projects represent large capital investments on behalf of hospitals that may take several years to complete. As a consequence, decisionmaking on compliance strategies occurs within the broader context of hospital business planning. In addition to the broader challenges to developing policy consensus noted at the beginning of this chapter, choices made by individual hospitals regarding retrofitting existing buildings, investing in new construction, or taking buildings out of service are contingent on multiple factors (see Table 4.4). A number of these factors, which emerged from discussions with hospital representatives, OSHPD, consulting engineers, and construction firms, are discussed below. Collectively, they reflect the complexity of the decisionmaking process that hospitals experience as they attempt to comply with SB 1953. These considerations may slow compliance, affect the costs of compliance, or influence the strategy that hospitals select to comply with.

Table 4.4. Key Factors Influencing Hospital Decisionmaking on SB 1953 Compliance Strategies

Decision Factor	Implications
Community needs	The needs of communities with respect to demand for different types of health care services influence long-term hospital business planning and therefore hospital seismic compliance strategies.
Compliance costs and benefits	The costs and benefits (to hospitals or communities) of seismic projects influence hospital seismic compliance strategies, as well as decisions regarding other types of hospital improvements.
Hospital finances	Hospital finances influence compliance strategies through their effect on the affordability of seismic upgrades, the ability to access financing and service debt, and the costs of capital.
Disruption of operations	Both structural and nonstructural retrofit projects can force hospitals to restrict operations and services while construction work is underway, resulting in loss of revenue and inconveniences for hospital workers and patients.
Cost escalation	The rate of cost escalation affects the total cost of a seismic project, particularly for complex projects that may take multiple years to complete.
Code requirements	Seismic retrofits of existing hospital buildings must be in compliance with current nonseismic building codes, which precludes hospitals from limiting seismic upgrades to only structural and nonstructural seismic deficiencies.
Land availability	In the absence of land for new construction, hospitals may be constrained in their options for seismic compliance. Those options include purchasing land at market value, demolishing and replacing existing buildings, or retrofitting existing buildings.
Time required for seismic project approval, permitting, and licensing	Through its effect on total project completion time, the administrative time associated with seismic projects exposes hospitals to cost escalation and could affect the ability of hospitals to complete necessary seismic upgrades before the 2030 deadline.
Uncertainty regarding future policy changes	Past changes in seismic compliance deadlines create policy uncertainty for hospitals and could create incentives for hospitals to delay action in anticipation of additional policy changes that could provide greater flexibility or regulatory relief in the future.

Community Needs

Hospital business planning and operations are conducted with consideration for the needs of the communities they serve. This is not only a general principle adopted by many hospitals but also a legal mandate under existing law. For example, the Patient Protection and Affordable Care Act of 2010 requires not-for-profit hospitals to conduct community health needs assessments (CHNAs) every three years.[37] This requirement was designed to enable hospitals to align the services they provide with the needs of communities. This emphasis on serving communities creates obligations for hospitals to consider the effects of SB 1953 compliance on their ability to meet identified community needs. For example, seismic construction projects that disrupt service provision, reduce the types of services that can be provided, or reduce the overall capacity of the hospital could have adverse consequences for communities. These consequences are likely to be more significant for those communities that have limited options in terms of hospital access, such as small, rural communities that may be served by a single hospital. To the extent that compliance with SB 1953 does interfere with meeting community needs, this could affect decisionmaking by hospital owners with respect to how and when they take action to achieve compliance.

Compliance Costs and Benefits

Although the needs of the community with respect to demand for health care services is the dominant factor driving hospital business planning and decisionmaking on investments in seismic upgrades, the resources available for such capital spending are finite. Therefore, in developing strategies for SB 1953 compliance, hospitals must weigh the costs of alternative compliance strategies (e.g., retrofit versus new construction). In so doing, hospitals seek to identify the lowest-cost strategy for achieving compliance, given other hospital objectives (e.g., minimizing disruptions to hospital operations, retirement of aging buildings, or increasing market competitiveness). Moreover, the question of cost is unavoidably linked to that of affordability, which is in turn a function of hospitals' overall financial situations (see "Hospital Finances," below). Accordingly, a significant proportion of this report is dedicated to the analysis of cost (Chapter 5) and affordability (Chapter 6).

Hospitals will also consider the potential benefits that could be generated in the process of pursuing seismic upgrades. The benefits of improvements that explicitly address seismic deficiencies arise only if a hospital experiences a seismic event of a magnitude that exceeds its existing structural tolerance. In the absence of such an event, the capital investment in enhanced seismic resilience is effectively a net loss for the hospital. In undertaking seismic compliance projects, however, hospitals are likely to pursue other types of nonseismic improvements. These may include mandatory improvements associated with compliance with other building codes (see

---

[37] Public Law 111-148, Patient Protection and Affordable Care Act, March 10, 2010. Affordable Care Act requirements for CHNAs are specified under Section 501(r)(3) of the tax code pertaining to tax-exempt charities and other 501(c)3 nonprofits. See Internal Revenue Service, "Community Health Needs Assessment for Charitable Hospital Organizations—Section 501(r)(3)," webpage, last updated November 7, 2018.

"Code Requirements," below), but they may also be voluntary improvements. For example, hospitals may choose to pursue actions to modernize the facilities and services at the same time as pursuing seismic upgrades, because the costs of doing so are less than if such modernization actions were pursued at a later date, independent of seismic work.

Hospital Finances

The hospital business model is wrought with complexity stemming from the mismatch between prices consumers pay for services and drugs and the reimbursement rates provided by insurance companies and government programs. As a result, hospitals make money on some services and lose money on other services. A variety of policy performance incentives and regulations exist that also guide decisionmaking, adding to business complexity. Additionally, hospitals face high fixed costs for building and technology capital and a variety of labor costs and challenges, both of which are dependent on the type of services hospitals choose to offer. Although far from a true competitive market, hospitals do face regional competition from each other, often on the basis of service quality and availability, because patients can choose where to obtain services and increasingly have access to comparative information on quality and cost. This can also influence hospital business decisionmaking and service provision.

A variety of hospital ownership models exist, which has direct implications on hospital business models and finances.[38] Private for-profit hospitals have a managerial incentive to focus on services that maximize profits and shareholder wealth. Private not-for-profit hospitals often have missions focused on community benefits or charitable services and often have religious affiliations. However, as private organizations, they also have an imperative to remain financially sustainable. Government hospital managerial incentives and organizational missions are similar to not-for-profit hospitals, but governmental hospitals often have an additional policy imperative to act as safety-net providers of services to low-income and uninsured populations. Private for-profit hospitals tend to have the highest operating margins, private not-for-profit hospitals have lower but positive operating margins, and public hospitals have the lowest operating margins, although many exceptions to these generalities exist.[39]

Despite differences in organizational goals and ownership types, all hospitals have the objectives of maintaining economic and organizational continuity, servicing debt, generating cash flows for future capital expansion, and providing high-quality patient care.[40] Managerial

---

[38] L. L. Liu, K. J. Jervis, M. M. Z. Younis, and D. A. Forgione, "Hospital Financial Distress, Recovery and Closure: Managerial Incentives and Political Costs," *Journal of Public Budgeting, Accounting and Financial Management*, Vol. 23, No. 1, 2011, pp. 31–68.

[39] One exogenous pressure that all hospital businesses face regardless of ownership type is the continuing changes to government reimbursement rates and policies to contain national health care costs. This has increased the number of financially distressed hospitals over time and is a trend that will likely continue (see Liu et al., 2011).

[40] Liu et al., 2011.

incentives often exist to reward administrators for meeting these objectives.[41] Because of increasing financial pressure and other strategic motives, the hospital landscape is currently in flux as administrators increasingly engage in mergers and acquisitions to improve cost efficiencies, offer additional services, and improve local market shares.[42]

Disruption of Operations

As noted in Chapter 5 (in the "Approach to Estimating Costs" section), seismic construction projects are rarely conducted on a blank canvas. Even when pursuing new construction, hospitals may need to relocate services and utilities, either temporarily or permanently, to accommodate the project. This preparatory, or "make ready," work is even more significant for retrofit projects.[43] Because of their age, SPC-2 buildings often lie at the heart of hospital campuses and can contain core departments, such as operating rooms, radiology, and intensive care units. Because these departments are important sources of revenue for hospitals and also are needed to support acute and critical care, hospital representatives frequently commented that they cannot simply be shut down while SPC or NPC work is undertaken.

Because of the need to maintain hospital operations and capacity while seismic work is being conducted, departments and services may be relocated to other parts of the campus—including into existing buildings or new buildings. Doing so may necessitate additional work to accommodate those services, and thus retrofit projects tend to have consequences beyond the noncompliant building. Hospital representatives indicated that, based on their experience, make-ready work invariably results in disruptions to hospital operations. This can include nuisance disruptions, such as noise, vibration, or restricted movement through the hospital. It could also include more-consequential effects, such as loss of revenue or capacity in the event a hospital has to scale back or temporarily cease certain operations or services.

Although structural retrofits to SPC-2 buildings can cause a range of disruptions, nonstructural work can cause significant disruptions as well. Because NPC requirements affect a larger number of hospitals and hospital buildings, NPC work is likely to be a more frequent source of disruptions. Although NPC work does not necessitate moving departments and health care services to new locations, the need to complete work in active areas of the hospital can nevertheless be a significant source of disruption. For example, completing NPC work in active operating rooms necessitates either taking individual rooms out of service while work is completed or working at night over extended periods. Both strategies have cost and capacity consequences.

---

[41] T. Flannery and D. Hofrichter, "What's the Right Incentive Compensation Plan?" *Trustee: The Journal for Hospital Governing Boards*, Vol. 60, No. 7, 2007, pp. 22–26.

[42] See Jacqueline LePointe, "How Hospital Merger and Acquisition Activity Is Changing Healthcare," Revcycle Intelligence, July 20, 2018.

[43] Make-ready work is composed of the movement of services, operations, and equipment around the hospital to allow access to areas requiring construction work.

Cost Escalation

During interviews, hospital executives raised concerns regarding the rising costs of construction in California—both in general and for hospitals in particular. Industry data suggest that construction costs have been rising faster than inflation over the past decade. General U.S. commercial construction costs since 2011 have averaged 4.1 percent per year,[44] compared with 2.3 percent for the consumer price index. This is consistent with one estimate of observed cost escalation from the 2007 RAND report of 4 percent per year.[45]

However, higher rates of increase in construction costs have been observed in California in recent years. For example, between 2016 and 2017, construction costs in Los Angeles and San Francisco increased by 7.1 percent and 7.6 percent, respectively.[46] Between 2017 and 2018, the increases were 4.8 percent and 6.2 percent, respectively.[47] These represent short-term rates of increase and could reflect transient cost volatility. Industry norms tend to be more modest. Nevertheless, given the complexity of hospital construction, projects may take multiple years to complete, with each additional year therefore adding additional costs. These effects are compounded by financing costs (see Chapter 5). Accounting for cost escalation in construction projects necessitates incurring additional debt to cover the additional cost.

Such cost escalation is a function of multiple factors. The active construction market in California creates competition for construction labor, materials, and contractors, particularly in the specialized market for hospital construction. Material costs can also fluctuate with competition and in response to broader commodity prices and trade policies. The large portfolio of projects that would need to be completed over the next decade to achieve compliance could potentially create a surge in construction demand and therefore greater competition for available contractors and labor. This could drive up the costs of construction, above what has been witnessed historically, and therefore the costs of compliance.

The anticipation of future cost increases could ultimately influence the type and timing of hospital capital investments. In principle, it incentivizes the completion of hospital projects over the near term rather than the long term. It also incentivizes the completion of projects as quickly as is practicable, assuming that funding and financing are adequate to complete the projects. This could also influence decisionmaking regarding the use of retrofit versus new construction to achieve compliance.

---

[44] Escalation rates are based on North American and U.S. construction costs from Rider Levett Bucknall. Quarterly reports are available at Rider Levett Bucknall, "Publications," webpage, undated.

[45] Meade and Kulick, 2007; the estimate is based on observed rates of increase in labor and materials.

[46] Rider Levett Bucknall, *North America Quarterly Construction Cost Report: Fourth Quarter 2017,* December 2017.

[47] Rider Levett Bucknall, *North America Quarterly Construction Cost Report: Fourth Quarter 2018,* December 2018.

Code Requirements

Because the life span of most building stock is long, compared with the frequency with which building codes are revised, it is common for a building not to be in accordance with the latest building codes. In most cases, building owners are not required to retrofit to meet current codes until they undertake a significant construction project. Such projects may include a major renovation or, in the context of this report, a retrofit to achieve seismic compliance. At that point, the building must be made compliant with all current codes.

Thus, when planning seismic projects to achieve compliance with SB 1953, hospitals must also bring all areas of the buildings affected by the project up to code. This includes codes from the Occupational Safety and Health Administration;[48] the Americans with Disabilities Act;[49] or the California Energy Code pertaining to patient safety, energy use, fire and life safety, and access for disabled persons.[50] The need to comply with these codes may constrain options for a seismic project. For example, between 2002 and 2012, the average square footage per hospital room increased by approximately 33 percent.[51] As hospitals update their facilities, this constraint could translate into a reduction in the number of rooms (and therefore beds) in a hospital building (see "Community Needs," above).

The fact that seismic retrofit projects trigger other California building code requirements largely impedes a hospital owner from undertaking improvements that are limited to addressing seismic deficiencies. Similarly, it means that the costs of construction for seismic compliance cannot be isolated from other types of additional construction work associated with a seismic project, because the two are inherently linked. This creates incentives for hospital owners to undertake other types of construction or make other types of upgrades and improvements (either in buildings' structure or facades or in equipment) when undertaking seismic work, assuming such additional work is affordable.

Land Availability

A common constraint identified by hospital representatives was limited access to available land for new construction. Hospital campuses have expanded as demand has grown, often keeping older, noncompliant buildings in service. Meanwhile, neighboring communities have grown as well. Although the issue is relatively clear in the dense urban environments of Los Angeles and the San Francisco Bay Area (e.g., San Francisco and Oakland), even hospitals in more-suburban settings may be effectively "land locked" by neighboring development or land

---

[48] See Occupational Safety and Health Administration, "Law and Regulations," webpage, undated.

[49] See U.S. Department of Justice, *2010 ADA Standards for Accessible Design*, Washington, D.C., September 15, 2010.

[50] California Energy Commission, "2019 Building Energy Efficiency Standards," webpage, undated.

[51] Esperanza Harper, Nicholas Watkins, and Tamra Minnier, "Patient Room Size Affects Satisfaction," Design Works HC, May 16, 2014.

ownership. Such circumstances may necessitate a range of responses on behalf of hospitals. These include the following:

- **Land acquisition:** If land is necessary for new construction, hospitals that do not already own land may have to negotiate the acquisition of new land, potentially at above-market rates (if there is high demand in the market). The costs of land acquisition become additional to the costs of construction, thereby driving up the costs of seismic compliance.
- **Demolition of existing buildings:** If new land acquisition is infeasible but new construction is the most cost-effective means of achieving compliance in alignment with hospital business planning, existing buildings or structures may have to be relocated or demolished to create room for construction. This could include demolition of parking facilities or demolition of the noncompliant buildings. This creates additional burdens for hospitals, however, because the costs of demolition or relocation have to be considered. In particular, if hospitals are demolishing SPC-2 buildings, the departments and services within those hospitals first have to be relocated to minimize service disruption (see "Disruption of Operations," above, and Chapter 5). This ultimately expands the total area of the hospital campus that is affected by seismic compliance and results in a range of potential indirect costs.
- **Forced retrofit:** In the absence of opportunities to acquire new land or rebuild in place, the options available to hospitals may be limited to retrofitting existing noncompliant buildings. The pursuit of this option assumes that the structural design of noncompliant buildings can support retrofitting. Although a seismic retrofit can be a cost-effective approach to achieving compliance, the resulting building may be less than optimal for serving the community. The retrofit may result in a reduction in hospital square footage and beds. Furthermore, working within the existing structure could preclude or constrain the ability to integrate new technologies or services, or at least drive up the cost of doing so.

The crowded nature of hospital campuses, combined with the expansion of outpatient services, is increasingly driving capital investment toward more-distributed outpatient facilities. Shifting demand to outpatient facilities, such as hospital-affiliated clinics and doctors' offices, can reduce the pressure on hospitals to maintain or enhance current capacity in the design and execution of seismic projects. Moreover, reimbursement rules allow hospitals to charge "facility fees" for services conducted at hospital-affiliated clinics.[52] Hence, the expansion of outpatient services and the acquisition of private practices by hospitals can have the dual benefits of reducing demand at inpatient facilities and enhancing revenue. As a result, although new construction on an existing hospital campus could potentially generate benefits for a hospital in terms of enhancing its competitive value, it might be preferable and under current rules to invest in outpatient facilities off campus.

---

[52] C. Capps, D. Dranove, and C. Ody, "The Effect of Hospital Acquisitions of Physician Practices on Prices and Spending," *Journal of Health Economics*, Vol. 59, 2018, pp. 139–152.

Time Required for Seismic Project Approval, Permitting, and Licensing

Completing seismic construction projects is a process that involves more than the act of construction alone. Part of this process involves working with OSHPD to obtain permits, monitor construction projects, and obtain compliance once the project is completed. Collectively, this process engages different components of OSHPD. First, hospitals apply for building permits by filing plans with OSHPD's Building Safety Section for review. Generally, the time it takes OSHPD to process construction plans and issue a permit varies significantly, particularly in response to the scale of the project. For example, projects valued at less than $50,000 are approved on an average of 90 days.[53] However, projects valued at greater than $10 million take an average of nearly a year to approve.

Once work commences, OSHPD field staff are responsible for monitoring projects to ensure that construction proceeds in accordance with the building code. Necessary changes to the plan that are encountered during construction may trigger a change order, which requires approval by OSHPD. As noted by hospital representatives, such approvals create the potential for construction delays.

Seismic projects also involve OSHPD's Seismic Compliance Unit (SCU), which is responsible for the review and approval of plans for achieving seismic compliance. This work consists of seismic evaluation reports, SPC ratings, compliance plans, and extension requests.[54] Therefore, seismic projects necessitate more planning and approvals than ordinary construction projects. This could include, for example, geotechnical reports or material testing reports that must be reviewed by the SCU. The SCU has the responsibility for approving completed projects and the reclassification of buildings. Completed projects also need licensing and certification by the California Department of Public Health before they can operate, and those applications require additional administrative time.

The direct administrative costs of compliance may be small relative to the costs of construction. Nevertheless, they are an important consideration, largely because of the time value of money. The greater the time between the start of a seismic project and when that project is licensed to operate and provide services, the greater the net cost to the hospital. Moreover, administrative processes have implications for the ability of hospitals to complete work prior to the 2030 deadline.

Uncertainty Regarding Future Policy Changes

As noted at the beginning of this chapter, formal policy efforts toward improving the seismic resilience of hospitals have been underway for nearly a half century—approximately the designed life span of a hospital building. During that period, there have been numerous policy changes—some occurring through legislative action and others through changes in rules

---

[53] The Plan Approval Date Estimator is available at OSHPD, "Plan Review Processes and Goals," webpage, undated.

[54] See OSHPD, "Application Process for the Seismic Compliance Unit," webpage, undated.

governing implementation. Therefore, any hospital weighing large capital investments to comply with seismic requirements is likely to consider the potential for future policy changes to emerge that might allow greater flexibility in achieving compliance (e.g., relaxed requirements or extended deadlines).

When such a potential exists, there can be value in delaying the investment—by postponing action or taking a more cautious, phased approach—until more certainty becomes available or until regulatory relief is offered. In other words, policy uncertainty can create financial incentives to delay compliance. On the other hand, such a delay tactic can be risky if greater flexibility fails to materialize. As a consequence, hospitals have taken different strategies with respect to how proactive they have been in achieving compliance. Some hospitals made early investments to replace aging buildings and campuses, others continue to work toward compliance with the 2020 and 2030 deadlines, and still others have yet to take any substantive action.

5. Estimates of SB 1953 Compliance Costs

One of the main concerns about the impact of SB 1953 is the cost that it imposes on California hospital owners. This chapter describes our approach to estimating costs and presents our best estimates for the statewide costs of complying with the bill going forward. Our estimates do not include the cost incurred by hospitals to date to comply with SB 1953. SB 1953 requires all general acute care hospitals to reach SPC-2 by 2020 (or by 2022 with an extension) and to reach SPC-4D and NPC-5 by 2030. Although there were still as many as 184 SPC-1 buildings in 76 hospitals as of September 2018, in most of these cases, a replacement building is underway or already completed and the SPC-1 building will be removed from general acute care service in 2030. Our cost estimate therefore addresses only the requirements for upgrading SPC-2 buildings to SPC-4D and for upgrading NPC-1 and NPC-2 buildings to NPC-5.[55]

Approach to Estimating Costs

Although earlier studies assumed that all hospitals would choose to replace noncompliant buildings with new buildings, this is far from the case, according to information that we have gathered through discussions with stakeholders and from cost estimates. Many hospitals are planning to retrofit existing buildings to bring them into compliance with SB 1953. We have therefore developed cost estimates for both approaches—retrofitting and replacing noncompliant buildings.

We initially sought to use OSHPD hospital construction project data to identify cost elements and their role in seismic upgrade projects. However, the OSHPD construction data are not suitable for this analysis for two main reasons. First, few buildings have been retrofitted from noncompliance to SPC-4D or SPC-5. This means that the vast majority of the retrofit costs in the OSHPD construction data are for retrofits from SPC-1 to SPC-2. These are distinct types of projects: Costs from one tell us little about costs for the other.

Second, the data do not contain enough information on seismic upgrade activities or costs. The available observations often do not indicate which buildings were involved in the upgrade. Moreover, the project descriptions are terse and do not describe the extent of projects well. Although hospitals voluntarily flag projects that include a seismic retrofit component, it is impossible to determine how much of the project involves seismic upgrading versus other activities and whether the project accomplishes all of the seismic upgrading needed.

[55] Although SPC-3 and SPC-4 buildings are compliant with SB 1953, these categories are for particular types of buildings. SPC-2 buildings cannot be upgraded to SPC-3 or SPC-4. They can be upgraded only to SPC-4D or SPC-5.

We therefore turned to discussions with hospital representatives to identify cost elements. Based on these discussions, we identified a number of general cost categories for hospital construction projects, such as preparatory or make-ready work;[56] construction soft costs;[57] and the construction work itself, which can be further broken down into specific elements. Beyond these direct costs, respondents also identified indirect impacts related to unpredictability, disruption, discontinuity of services, and possible temporary loss of capacity, all potentially leading to decreased revenue.

The relevance of these general cost elements emerged reasonably consistently from discussions with hospital representatives, but we were less successful in determining how to apply them to hospital seismic upgrade projects. In principal, building characteristics, such as age and construction type, could be used to determine the type and extent of work needed for SPC and NPC upgrades. In addition, spaces devoted to different types of services—such as patient rooms, surgery, labs, or dietary services—presumably have different costs. However, hospital representatives emphasized that the complexity and individualized nature of hospital designs, settings, and circumstances present substantial impediments to any sort of systematic building-block approach to estimating seismic upgrading costs.

We therefore roll all cost elements into two discrete upgrade options: retrofit or replace noncompliant buildings. Costs for each were taken from a sample of hospitals and extrapolated to the entire population of hospitals needing upgrades. Given the individualized nature of the upgrade projects, this approach is less accurate than building detailed estimates for each hospital from specific cost elements. As desirable as such estimates are, they are not feasible with the information available.

Data and Methods

Data Sources

The data sources we used in developing our cost estimates are summarized in Table 5.1. These include data from OSHPD on seismic compliance status and numbers of hospitals, buildings, and beds, as well as market data on costs of seismic retrofitting and new construction. The hospital seismic compliance data set provides SPC and NPC ratings by building for all buildings subject to compliance with SB 1953. The SPC-2 buildings data set contains the number of beds in SPC-2 buildings for 249 hospitals, as estimated by OSHPD field staff familiar with the individual hospitals. This estimate has not been verified by the hospitals themselves.

The final data set in Table 5.1 includes actual costs for completed replacement hospitals provided by construction firms, retrofit and replacement cost proposals developed by

---

[56] This means moving services and operations around the hospital to allow access to areas requiring construction work.

[57] These are construction-related costs, excluding direct construction labor and materials, such as architecture, engineering, consultants, inspections and testing, permits and fees, furnishings and equipment, information technology, and project management.

construction firms for hospitals, and retrofit and replacement cost estimates developed by hospitals themselves. The data comprise 147 buildings from 68 hospitals that collectively contain more than 10,000 beds. The sample of hospitals for which we obtained cost data is reasonably representative of California hospitals overall. The hospitals range from some of the largest in the state to some having fewer than 50 beds. Figure 5.1 shows comparisons of the sampled hospitals with the rest of California hospitals in terms of number of beds, number of buildings, and total number of patient-days. These are indicators of hospital size, and the distribution is similar across the sampled and nonsampled hospitals. The sampled hospitals are distributed geographically—26 of the 58 California counties contain at least one of the sampled hospitals. Although construction costs are known to vary with geographic location, such variation is small compared with the uncertainties in our cost data. We therefore do not attempt to explicitly account for this variation. By drawing our sample from such a wide geographic range, we nevertheless expect that geographic cost variability is accounted for in our average cost estimates.

Table 5.1. Data and Sources Used to Estimate Costs

Data Set	Primary Data Fields Used	Number of Samples	Data Source
Hospital building seismic compliance status, September 2018	SPC and NPC rating	3,632 buildings	OSHPD, "Hospital Building Data," data set, accessed September 2018
Bed counts in SPC-2 buildings, January 2018	Number of beds in SPC-2 buildings	638 buildings	Data provided to RAND by OSHPD
Seismic retrofit and new construction cost estimates, by building	Cost, number of beds	• 87 NPC retrofit estimates • 49 SPC retrofit estimates • 54 new construction estimates • 68 hospitals	Data provided to RAND by construction firms and hospitals (construction firm bids and in-house estimates)

Figure 5.1. Comparison of Sampled Hospitals' Retrofit Costs with the Rest of California Hospitals, by Attribute

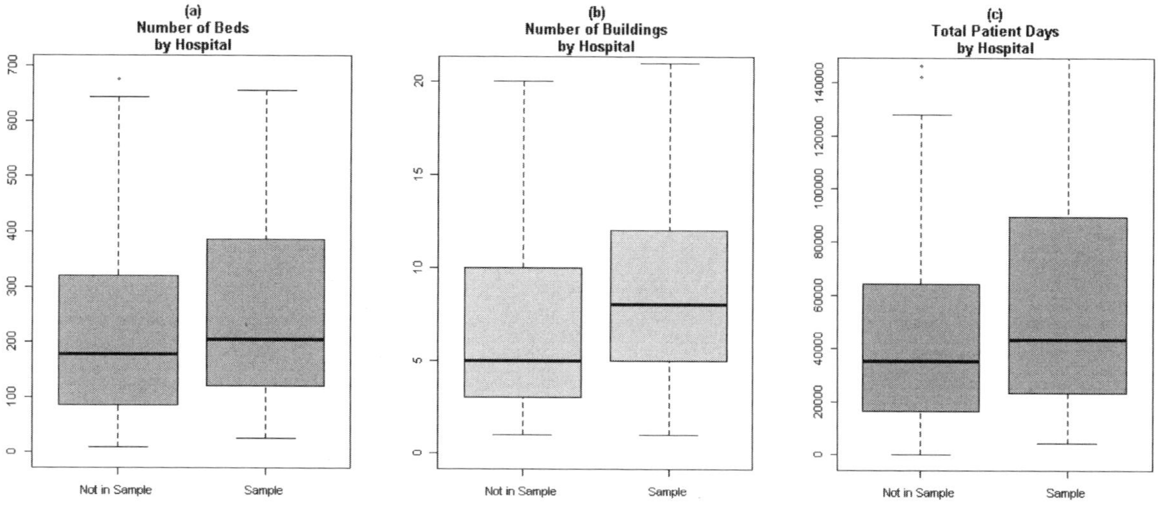

NOTES: The solid bar shows the median values, the top and bottom of the boxes show the 25th and 75th percentile values, and the whisker ends extend 1.5 × box height (i.e., the difference between the 75th and 25th percentiles) with maximum and minimum data points as bounds. See Table 5.1 for information about data sources.

Adjustments to Cost Data

The SPC retrofit cost estimates are for upgrading SPC-2 buildings to SPC-4D. The NPC retrofit cost estimates are generally to upgrade NPC-2 buildings to NPC-5. In some cases they are for upgrades to NPC-3 only. Upgrading from NPC-3 to NPC-4D level 1 requires only the addition of an emergency plan, for which we assume the cost is negligible. The only construction required to bring an NPC-3 building into compliance is to add the fuel and water storage systems required for NPC-5. We therefore augment the NPC-3 retrofit cost estimates with cost of these storage systems. We approximate the NPC-5 retrofit cost as $2 million per hospital campus, which is the typical value cited in the more-detailed cost estimates in our sample. This NPC-5 cost is distributed equally among all buildings within a hospital.

The retrofit and replacement cost estimates include preparatory make-ready work and generally also include soft costs. In the cases where soft costs were explicitly excluded from the estimate (4 retrofit and 13 replacement estimates), we augment the estimates to include them. We approximate the soft costs as 60 percent of the construction costs. This factor is derived from a subset of detailed cost estimates for which soft costs are explicitly characterized; it was further cited independently or corroborated on three separate discussions with health care construction experts.

Finally, most cost estimates we received were for future work, and most were escalated to account for increasing costs in future years. Similarly, costs from completed projects were in prior-year dollars. To make costs from different years comparable, we converted all costs to 2019 dollars using a 4 percent annual escalation rate. This value is typical of the values used in the cost estimates to escalate the costs and is also consistent with the discussion in the "Cost Escalation" section in Chapter 4.

Building Retrofit and Replacement Costs

Cost data are summarized in Tables 5.2–5.4. The conventional way to present construction costs is in terms of cost per area. Indeed, in cases for which we have area values, cost correlates well with building area. Unfortunately, we do not have data on building area for most California hospitals.

In lieu of area, we normalized our cost data to the number of beds. Number of beds is a common proxy for hospital size, but it is an imperfect metric for construction costs. The size of rooms can vary, as can the area devoted to such requirements as emergency departments, operating rooms, diagnostic and test facilities, dietary services, utility plants, and other non-bed space. Also, because construction projects must comply with nonseismic-related code requirements, such as access for disabled persons, seismic upgrading sometimes leads to larger and fewer rooms. And, often, hospital owners choose to change the size and number of beds or rooms for other reasons. For all these reasons, cost per bed is a nonideal cost metric. In addition, nearly half the SPC-2 hospital buildings in California have no beds, in which case cost per bed has no meaning. As a result, we use both cost per building and cost per bed in our estimates.

The SPC retrofit cost data are summarized in Table 5.2. The cost per building spans a substantial range, and the average cost for all buildings combined masks some important variability among buildings. Most important, costs for buildings with beds are substantially higher than costs for buildings without beds. This difference primarily reflects building size, because beds are typically located in large buildings labeled as "main" or "tower," while buildings without beds are generally smaller and often house more-specific functions, including utility plants, specialty services, and administrative functions.

For buildings with beds, we also present the cost per bed. Even after normalizing costs by the number of beds, there is substantial variability. This variability stems from the wide range in the extent and complexity of retrofit requirements, which, in turn, stem from the wide range in building design, age, relationships with adjacent buildings, and numerous other idiosyncratic details. Such variation exemplifies the challenge with estimating costs for complying with SB 1953.

Table 5.2. SPC-2 to SPC-4D Retrofit Costs from Sample Data

		Cost per Building			Cost per Bed		
Building Group	Number in Sample	5th Percentile	Average	95th Percentile	5th Percentile	Average	95th Percentile
All	49	$768,451	$92,827,296	$404,293,678			
With beds	35	$1,798,885	$128,512,120	$597,867,788	$51,684	$462,155	$1,853,037
Without beds	14	$507,277	$3,615,237	$10,334,482			

NOTE: See Table 5.1 (row 3) for information about data sources.

In the case of NPC retrofits (Table 5.3), cost estimates could not be distinguished by whether a building has beds or not because bed counts are available only for SPC-2 buildings (see Table 4.2). Although many NPC retrofit cost estimates apply to SPC-2 buildings, many others apply to buildings that are SPC-3 to SPC-5 (SPC >2), for which we do not have bed counts. We find that NPC retrofit costs for SPC-2 buildings are substantially greater than costs for SPC >2 buildings. This presumably reflects the fact that SPC-2 buildings are older than SPC >2 buildings and hence have greater NPC retrofit requirements. We also find that, for SPC-2 buildings, NPC retrofit costs correlate reasonably well with SPC retrofit costs. When estimates of both costs are available for a building group, Table 5.3 presents the ratio of NPC retrofit costs to SPC retrofit costs. This allowed us to link NPC retrofit costs to SPC retrofit costs.

Table 5.3. NPC-2 to NPC-5 Retrofit Costs from Sample Data

		Cost per Building			NPC Cost/SPC Cost		
Building Group	Number in Sample	5th Percentile	Average	95th Percentile	5th Percentile	Average	95th Percentile
All	87	$405,389	$10,355,986	$32,453,366	—	—	—
SPC-2	45	$385,035	$15,505,641	$53,364,736	0.051	0.50	1.08
SPC >2	42	$702,271	$4,838,497	$16,600,000	—	—	—

NOTE: See Table 5.1 (row 3) for information about data sources.

The new cost data for hospital construction are shown in Table 5.4. As with the SPC retrofit costs, there is a substantial difference in new construction costs between buildings with and without beds, which we again ascribe primarily to differences in the sizes of these two types of buildings. Although a range in the cost per bed is again apparent, the range is much less than that for the SPC retrofit cost per bed (Table 5.2). This is consistent with new construction requirements and costs being more uniform than those for retrofitting.

Our new construction cost estimates are based on a combination of actual costs for recently completed buildings (within the past decade) and estimates for planned construction of new buildings. As an internal check on our data, we compared cost estimates for recently constructed buildings with those for planned construction. The average costs per bed for the two groups are within 20 percent of each other, indicating that estimates for future work are reasonably consistent with actual costs from recently completed hospitals.

Table 5.4. New Construction Costs for Hospitals from Sample Data

Building Group	Number in Sample	Cost per Building: 5th Percentile	Cost per Building: Average	Cost per Building: 95th Percentile	Cost per Bed: 5th Percentile	Cost per Bed: Average	Cost per Bed: 95th Percentile
All	54	$5,247,084	$479,072,905	$1,324,323,957			
With beds							
All	43	$10,656,697	$588,465,004	$1,389,271,175	$800,526	$3,068,767	$6,685,466
Completed	10	$167,478,766	$568,289,885	$1,347,941,127	$1,726,004	$2,567,163	$3,859,890
Planned	33	$7,737,798	$594,578,677	$1,400,410,240	$680,000	$3,220,769	$7,168,802
Without beds	11	$3,709,295	$51,449,244	$138,418,370			

NOTE: See Table 5.1 (row 3) for information about data sources.

Statewide Compliance Cost Estimates

We use the cost data described above to estimate statewide costs for complying with the SB 1953 requirements of upgrading SPC-2 buildings to SPC-4D and of upgrading NPC-1 and NPC-2 buildings to NPC-5. As described above, decisions about major capital projects are complex, and the approach to complying with SB 1953 compliance is inexorably tied up with several other considerations. As a result, the decision about whether a hospital will choose to retrofit, replace, or some combination of the two is impossible to determine a priori. Representatives for many hospitals we met with noted that they are still engaged in an extended decisionmaking process and have yet to develop upgrade plans.

In light of this uncertainty, we develop cost estimates for two end-member cases: a lower bound in which hospitals choose to retrofit all noncompliant buildings and an upper bound in which hospitals choose to replace them. These end-member cases are then used to compute costs for intermediate cases, in which different proportions of hospitals choose to retrofit and replace.

Construction cost estimates are conventionally reported as escalated costs. Cost escalation is the practice of increasing costs that will be incurred in the future to account for the increase in costs due to inflation and other factors. To escalate our cost estimates, we assume that SB 1953–related construction projects will be evenly distributed in time between 2019 and 2030, such that we can treat the statewide total cost as a single project running from 2019 to 2030. Based on the escalation values used in the cost data provided by hospitals, our discussions with construction firms, and the discussion in the "Cost Escalation" section in Chapter 4, we assume an escalation rate of 4 percent per year. Some stakeholders noted that the fact that all hospitals in California must complete their construction work by the 2030 deadline could create a surge in demand for construction services that would create a supply shortage and concomitant increase in costs. This sort of demand surge, well-known in the insurance industry, is generally encountered after large-scale disasters in which many properties are damaged. Although health care construction is a specialized niche in the construction market, it is unclear whether the 2030 compliance deadline for SB 1953 in California would create substantial demand surge. To account for this possibility, we also present escalated costs computed using an escalation rate of 4 percent from 2019 to 2024 and 8 percent from 2025 to 2030.

Retrofit Costs

SPC and NPC retrofit costs are extrapolated statewide in Table 5.5. Each row of the table shows a particular group of buildings requiring seismic upgrading and how costs for that building group are estimated. Of the 631 SPC-2 buildings (Table 4.3), 322 have beds, and costs are estimated using the average cost per bed figure from the sample data in Table 5.2. For the 300 SPC-2 buildings with no beds and 9 with no data on beds, costs are estimated with the average cost of buildings without beds from the sample data in Table 5.2. The total unescalated SPC retrofit cost estimate is $14 billion.

Table 5.5. Statewide Retrofit Costs

Retrofit Type	Building Group	Buildings in Group	Beds in Group	Retrofit Cost per Bed	Retrofit Cost per Building	Total Cost	Average Cost per Building
SPC	SPC-2 with beds	322	27,255	$462,155	—	$12,596,034,525	$39,118,120
SPC	SPC-2 with no beds	309	—	—	$3,615,237	$1,117,108,233	$3,615,237
SPC	Total SPC costs	631	—	—	—	$13,713,142,758	$21,732,397
NPC	NPC-2 *and* SPC-2	519	—	—	0.50 × SPC retrofit cost	$5,488,534,599	$10,575,211
NPC	NPC-2 *and* SPC >2	1,148	—	—	$4,838,497	$5,554,594,556	$4,838,497
NPC	NPC-1 *and* SPC >1	109	—	—	$4,838,497	$527,396,173	$4,838,497
NPC	Total NPC costs	1,776	—	—	—	$11,570,525,328	$6,514,935
Total cost	—	—	—	—	—	$25,283,668,086	—
Escalated total, 4%/year	—	—	—	—	—	$31,658,956,507	—
Demand surge, escalated total for 2019–2024 = 4%/year; 2025–2030 = 8%/year	—	—	—	—	—	$34,285,237,904	—

NOTES: Cost totals neglect an estimated $250 million for NPC-5 retrofits to NPC-3 and NPC-4 buildings. — = not applicable. See Table 5.1 for information about data sources.

Of the 1,810 NPC-2 buildings (Table 4.3), 519 are SPC-2, and costs are derived from the SPC retrofit cost for that building using the ratio of NPC and SPC retrofit costs of 0.5 (Table 5.3). For the 1,148 SPC >2 buildings, we estimated costs using the average NPC-2 retrofit cost for SPC >2 buildings from the sample data. Table 5.5 also includes a retrofit cost estimate for 109 NPC-1 buildings that are SPC >1. We have no cost estimates for upgrades of NPC-1 buildings, so we again use the average NPC-2 upgrade cost for SPC >2 buildings from the sample data. While this may underestimate costs for upgrades from NPC-1 to NPC-5, it accounts for at least part of the upgrade cost for these NPC-1 buildings. The total unescalated NPC retrofit cost estimate is $12 billion.

Summing SPC and NPC retrofit costs gives a total unescalated retrofit cost of $25 billion. Applying a 4 percent per year cost escalation brings the total to $32 billion, and doubling the escalation rate from 2025 to 2030 brings the total to $34 billion.

Replacement Costs

New costs for hospital building construction are extrapolated statewide in Table 5.6. This estimate includes the cost of replacing all SPC-2 buildings and completing NPC retrofits on all SPC >2 buildings. Note that this estimate neglects some important factors, some of which are discussed in Chapter 4, that may influence costs. These include the costs of repurposing or demolishing or, conversely, the revenue from selling; the costs of buildings being replaced; the costs of acquiring land for replacement buildings when it is not already available; and the costs of improvements to nearby public infrastructure imposed by cities as a condition of granting construction permits.

Table 5.6. Statewide Replacement Costs

Building Group	Buildings in Group	Beds in Group	Cost per Bed	Cost per Building	Total Cost	Average Cost per Building
SPC-2 with beds	322	27,255	$3,068,767	—	$83,639,254,881	$259,749,239
SPC-2 with no beds	309	—	—	$51,449,244	$15,897,816,246	$51,449,244
Total SPC-2 replacement	631	—	—	—	$99,537,071,127	$157,744,962
SPC >2 and NPC-1 or 2 (retrofit)	1,220	—	—	$4,838,497	$5,902,966,340	$4,838,497
Total cost	—	—	—	—	$105,440,037,467	—
Escalated total 4%/year	—	—	—	—	$132,026,790,927	—
Demand surge, escalated total for 2019–2024 = 4%/year; 2025–2030 = 8%/year	—	—	—	—	$142,979,126,164	—

NOTES: — = not applicable. See Table 5.1 for information about data sources.

Just as with the SPC retrofit costs, the cost for the 322 SPC-2 buildings with beds is estimated using the average replacement cost per bed, while the cost for the 300 buildings with no beds and 9 buildings with unknown beds is estimated from the average replacement cost of buildings without beds. The total unescalated cost of replacing SPC-2 buildings comes to just under $100 billion, or an average of $158 million per building. We add to this the NPC retrofit costs for SPC >2 buildings. As was done for the retrofit cost estimate, we use the cost to retrofit an NPC-2 building for both the NPC-1 and NPC-2 buildings. NPC retrofits to 1,220 buildings add almost $6 billion, bringing the total unescalated cost to $105 billion. Applying a 4 percent per year cost escalation brings the total to $132 billion, and doubling the escalation rate from 2025 to 2030 brings the total to $143 billion.

Given that the cost estimates for building replacement were self-reported by hospitals, a question that arises is whether an independent validation of these costs is possible. As a first-order check, we recalculated the replacement cost using the number of SPC-2 beds, some common rules of thumb, and industry data on construction costs. One common rule of thumb is that modern hospitals are now built to allocate up to approximately 2,500 square feet per bed. Given 27,255 SPC-2 beds (as in Table 5.6), a total area of 68,137,500 square feet of SPC-2 floor space would need to be replaced. One recent industry report estimated the cost per square foot of hospital construction in Los Angeles and San Francisco for the fourth quarter of 2018 at $675–$780.[58] At that unit cost, the total replacement cost is estimated at $46.0 billion–$53.1 billion without accounting for escalation. Adding soft costs (again assuming 60 percent of the construction cost) brings this total to $73.6 billion–$85.0 billion. The upper end of this range is within 2 percent of the $83.6 billion total cost for replacing only those SPC-2 buildings with beds based on hospital self-reported project costs (Table 5.6). Hence, the costs for replacement in Table 5.6 appear consistent with what one might expect, independent of the project costs that hospitals themselves have estimated.

Total Cost of Compliance

Using these two end-member approaches to compliance—entirely through retrofitting and entirely through replacement—we can examine total costs in terms of the relative proportions of these two approaches used. Figure 5.2 uses a simple linear interpolation to show the statewide escalated total cost of compliance as a function of the proportion of compliance achieved through replacement. For example, if 30 percent of compliance was achieved through replacement, we estimate that the total cost of statewide compliance would be approximately $62 billion. As discussed above, we are unable to determine how many hospitals will use one approach or the other and therefore cannot determine where along the curve in Figure 5.2 the true cost lies.

---

[58] Rider Levett Bucknall, 2018.

Figure 5.2. Statewide Compliance Costs as a Function of Proportion of Compliance Achieved Through Replacement

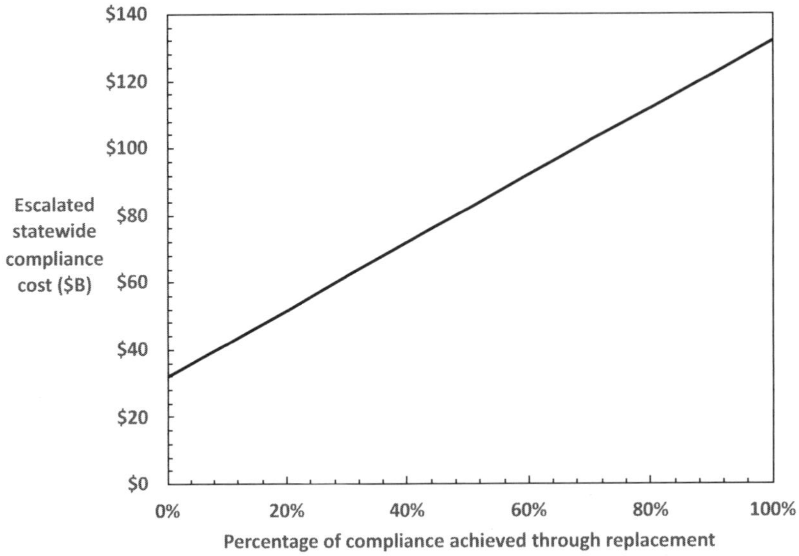

Financing Costs

An important aspect of large capital investments, such as upgrading building infrastructure, is obtaining the capital required for the project. Although some hospitals may be able to pay for such projects through a combination of operating revenues, endowments, and philanthropy (nonprofit) or tax revenues (public), given the size of the projects being undertaken and the precarious state of many hospitals' financial situations (see next chapter), we expect that most hospitals would need to take on new debt to finance seismic upgrade projects. The extent to which hospitals would finance a portion or all of the overall project costs is highly uncertain and dependent on individual hospital circumstances, so we do not attempt to estimate this parameter.

The cost of obtaining capital can be substantial. To provide a sense of this potential cost, Figure 5.3 presents some simple hypothetical computations of the cost of financing $20 billion for different payback periods and interest rates. These results show that financing costs could increase the cost of the project by a factor of two or more.

Figure 5.3. Potential Financing Costs on $20 Billion

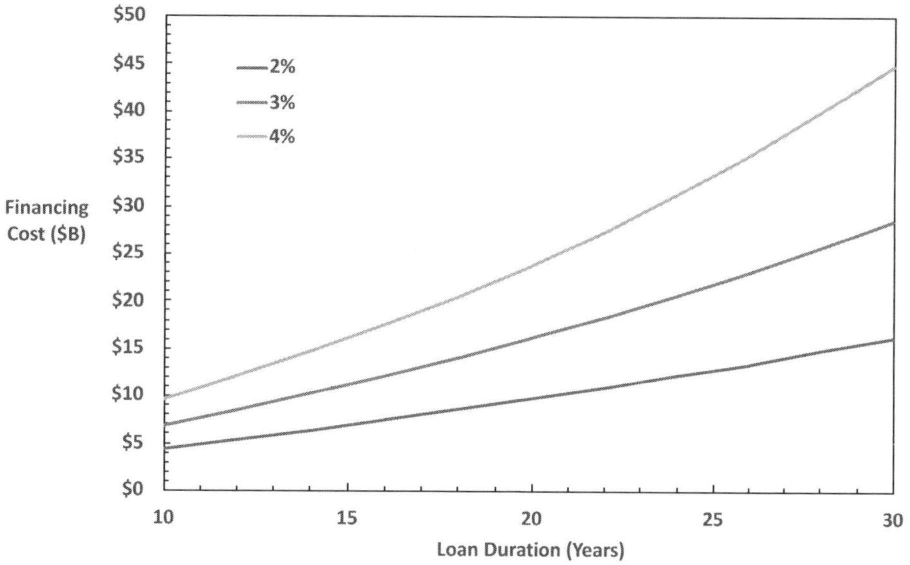

Assumptions in Cost Estimates

We have endeavored to capture the primary costs involved in complying with SB 1953, but our cost computations necessarily rely on some assumptions that limit the accuracy of our estimates. Table 5.7 lists these assumptions and the sense in which they affect our cost estimates. The first four assumptions highlight potential costs that we have been unable to quantify. Some, such as potential revenue losses resulting from construction activities, could be substantial. The fifth and sixth highlight assumptions we have used in extrapolating our costs that are necessitated by limitations in the type and amount of data available to us: Information about the areas of hospital buildings in California would eliminate the need for these assumptions. The seventh assumption refers to neglecting cases in which hospitals are unable to make the investments needed to comply with SB 1953 and are compelled to close buildings rather than attain compliance. This would eliminate upgrading costs and possibly also generate revenue from sales. While we have heard of no definitive plans to pursue this path, it remains a possibility in some cases.

Table 5.7. Costing Assumptions and Their Effect on Cost Estimates

Assumption	Effect on Cost Estimates
1. Neglect revenue loss or operational cost increases caused by construction disruption	Underestimates actual
2. Neglect cost of acquiring land for expansion or replacement buildings	Underestimates actual
3. Neglect cost of repurposing or demolishing buildings being replaced	Underestimates actual
4. Neglect cost of developing emergency plans	Underestimates actual
5. Cost per bed is an accurate cost measure for buildings with beds, and the ratio of bed space to non-bed space in all hospitals is equal to that in the hospitals from which our sample data were derived	Uncertain
6. Cost per building is an accurate cost measure for buildings without beds, and the cost per building in all hospitals is equal to that in our sample data	Uncertain
7. Neglect cases where buildings will not be retrofitted or replaced but rather removed from general acute service, closed, or sold	Overestimates actual

Costs in Context

If hospitals are to meet the 2030 deadline and construction progresses steadily between now and then, our cost estimates imply statewide annual hospital construction spending starting at $2 billion (retrofitting) to $9 billion (replacement) per year in 2019 and increasing from there because of cost escalation. For comparison, OSHPD construction data show that California hospitals spent $44.5 billion on construction between 2000 and 2018, or an average of $2.3 billion per year.[59] The spending rate for our lower-bound cost falls below this spending rate, although it is more than four times this rate for the upper bound. It is unknown whether the supply of health care construction firms can meet this demand. To the extent that demand increases over the historical average, the market may experience demand surge pricing, meaning that costs could increase more than normal escalation because of a shortage of supply.

Seismic upgrading projects are major capital investments, and, as noted earlier, most projects will include—either voluntarily or mandatorily—nonseismic-related improvements, such as repair and maintenance, required nonseismic code upgrades, and changes reflecting current hospital design (e.g., larger rooms, converting from shared to private rooms, ensuring rooms have windows, larger diagnostic and test facilities). In this perspective, some portion of costs for projects nominally attributed to seismic compliance is actually part of a background trend of (nonseismic) capital reinvestment spending to offset depreciation and design obsolescence. To the extent that this ongoing capital reinvestment work is included in seismic upgrade projects, then the actual "seismic" portion of the project costs would be less than the compliance cost estimates presented in this chapter. We can get a rough idea of the magnitude of this background capital reinvestment spending from the OSHPD construction data. Of the $44.5 billion in

[59] This is based on construction project data supplied to RAND by OSHPD.

construction spending since 2000, $41.8 billion, or $2.2 billion per year, was not flagged as seismic upgrade–related work and could be considered ongoing capital reinvestment spending. The portion of this work that may be included in seismic retrofit projects is unclear. As an upper bound, if seismic upgrade projects include all the capital reinvestment work that hospitals would otherwise undertake between now and 2030, compliance costs could be as much as $26 billion less than those presented in this chapter.

RAND's 2007 report estimated a cost range of $59 billion to $145 billion (in 2019 dollars; see Table 2.4).[60] That report addressed SPC-1 building space and assumed that all hospitals would attain compliance through building replacement. Consequently, that range is nominally comparable to our replacement cost estimate of $105 billion. Direct comparison of the two estimates is complicated by changes over time in the amount of the hospital infrastructure that needs to be brought into compliance and differences in the analytical approaches used. At the time of that earlier report, SPC-1 buildings accounted for approximately 41.5 million square feet and 34,850 beds—greater than the 27,255 beds in SPC-2 buildings used in this analysis. The earlier report also assumed a maximum construction rate per year that forced construction to run for about 45 years at an annual escalation rate of 2 percent, significantly increasing the escalation factor relative to the 12 years assumed in this estimate. To facilitate comparison, we recompute our escalated estimates using an inflation rate of 2 percent for 45 years. This procedure yields an escalated total cost of $167 billion, which is greater than the upper bound from the 2007 report.

---

[60] Meade and Kulick, 2007.

6. Affordability of SB 1953 Compliance for California Hospitals

Policies should balance seismic safety benefits with potential unintended adverse effects to hospital finances, capacity, and patient populations. SB 1953 compliance would increase the hospital's resilience to earthquakes and reduce the risk of earthquake damage immediately reducing hospital capacity and services offered in the wake of a major seismic event. However, the financial stress imposed by its requirements could also force managerial decisions that may also result in hospital capacity reduction. It is beyond the scope of this work to conduct a cost-benefit analysis of the SB 1953 requirements in which finances and effects on human life are considered, but this chapter provides an assessment of the potential magnitude and distribution of SB 1953–imposed financial distress on hospitals and by hospital type, an important element to consider when assessing seismic policies.

We define *affordability* as the capability of hospitals to comply with SB 1953 given their current financial health. In the future, when many hospitals undertake retrofits, actual affordability will also depend on business model, operational, and financial changes that hospitals undertake in the interim; however, such changes are impossible to predict and are beyond the scope of our analysis. *Affordability* as we define it for this report is most directly related to the quality of a hospital's current balance sheet (assets and liabilities) and recent operating performance (profit/loss), but it is also indirectly related to health care market trends and other important characteristics, such as size, location, and ownership structure.

When taking into account these factors of affordability, no hospital is exactly alike, and the sector collectively exhibits a wide range of financial health conditions. Some hospitals are extremely well endowed and have high profit margins, others have moderate financial health, and some exhibit severe distress. Financial distress presents a significant challenge to hospital administrators and can force a variety of managerial decisions,[61] including but not limited to

- spinning off or divesting assets
- increasing debt offerings
- merging with other hospitals or acquisition by an MHS (independent hospitals only)
- appealing for additional government funds (public hospitals only)
- changing management
- restructuring service lines to improve profitability
- decreasing the capacity of or closing service lines that are unprofitable
- diversifying into new, more-complex service lines with greater acuity
- improving occupancy rates by improving referral networks and market visibility

[61] This list was adapted from the following sources: Liu et al., 2011; J. Langabeer, "Predicting Financial Distress in Teaching Hospitals," *Journal of Health Care Finance*, Vol. 33, No. 2, 2006, pp. 84–92; James R. Langabeer II, Karima H. Lalani, Tiffany Champagne-Langabeer, and Jeffrey R. Helton, "Predicting Financial Distress in Acute Care Hospitals," *Hospital Topics*, Vol. 96, No. 3, 2018; and interviews with hospital representatives and other experts.

- shifting toward a greater mix of outpatient services
- changing the coding of services—i.e., upcoding—to receive higher reimbursement rates
- delaying capital investments in buildings or technology
- closing hospitals.

In terms of patient care, such changes could have an adverse impact on populations seeking unprofitable services, no impact, or an indirect but beneficial impact on services by improving the financial resiliency of the hospital. All options are not necessarily available to all types of hospitals, and several options' ability to improve financial health can be exhausted; also, a hospital may inevitably face a situation of constrained decisionmaking that has adverse impacts on communities or specific populations. In the case of SB 1953 compliance, it is impossible to forecast what business decisions hospitals may make as a result of further or induced financial distress from a mandatory major capital investment for two reasons: (1) Research indicates that there is little guide to predicting such decisions for any hospital, which hospital administrators also corroborated, and (2) there is little to no precedent for the magnitude of the financial impact that SB 1953 imposes on California hospitals.

Our analysis indicates that hospitals in severe financial distress might not be able to or might be severely constrained in their ability to afford to implement actions to meet the SB 1953's 2030 seismic compliance requirements. In interviews, hospital administrators emphasized that if forced to implement infrastructure projects they cannot afford, they will consider the actions listed above to make such requirements financially feasible and additionally may remove general acute care services from noncompliant buildings to avoid seismic retrofitting cost burdens, potentially reducing hospital capacity. Closure, reduction of capacity, or reduction of service lines will adversely affect communities that rely on such services, a potential trade-off that policymakers should consider as result of SB 1953 compliance. In the event of such an outcome, the financial impact of SB 1953 would unlikely be the sole causal reason for the decision but rather one of several preceding factors causing financial distress.

It is difficult to assess exactly how the seismic requirements will affect California hospital finances. However, there is published research that provides methods for categorizing the financial health of organizations; in recent years, these methods have been applied to general acute care hospitals elsewhere. We do not use these methods to identify the financial health of individual, specific hospitals. Instead, we characterize the general trend of financial distress across hospital types and created by the seismic requirements. Such characterization can inform current and future policy assessment of hospital seismic requirements in an effort to balance seismic safety, affordability, and adverse effects on the populations served.

Methods for Characterizing Hospital Financial Distress

There are several ways to assess hospital financial characteristics. Hospital executives often are concerned with key financial and operating statistics, including operating margin, current ratio, debt service coverage ratio, average age of plant, occupancy rates, and days cash on hand. Each of these provides performance information about a different element of a hospital's business model that executives may use to inform decisionmaking. However, from an external research perspective, it is difficult to accurately and objectively assess how a combination of these metrics interacts to influence holistic financial health and which metrics are relatively more important than others. In addition, it is difficult to establish objective thresholds within these data to determine where a hospital lies in the financial health spectrum or whether it is comparatively better or worse off than a different hospital.

One method of assessing financial health is through ratings provided by credit rating agencies, but this method is difficult to apply to California's general acute care hospitals. These metrics provide investors an accurate projection of a hospital's likelihood of meeting debt obligations and are often a good proxy for financial health. However, many hospitals in California do not have credit ratings because they have not recently sought debt financing from capital markets, and many of the hospitals lacking credit ratings tend to be at the middle to lower end of the financial health spectrum that we discuss establishing below.[62] Credit ratings for hospitals in large MHSs and publicly owned hospitals might not provide a holistic picture of financial health because the ratings are based on the fiscal strength of the owner (e.g., Los Angeles County) rather than the hospital itself.

Instead, we chose to use to the Altman Z score as a method for categorizing financial health.[63] The Altman Z score is a widely used and successful method for categorizing the financial health of an organization.[64] The original Altman Z score is a discriminant analysis regression model of financial ratio variables originally developed in 1968 to predict corporate bankruptcies. Since then, researchers have created variants of the Z score and have tested and successfully applied these variants to characterize corporate financial distress in other industries, including emerging markets and nonmanufacturing industries.[65] The Altman Z model developed for nonmanufacturing industries is the most appropriate Z score variant for acute care hospitals. It has previously been used in three studies—to assess financial distress among 50 teaching hospitals,[66] a national sample of more than 600 public hospitals,[67] and 310 acute care hospitals in Texas, including public, private for-profit, and private not-for-profit hospitals.[68] Equation 6.1 displays the Altman Z score model:

---

[62] We estimate that approximately 75 percent of California hospitals currently have credit ratings from Moody's, Fitch, or S&P. It is normal for fiscal entities below a certain size to not have credit ratings from big rating agencies.
[63] Edward I. Altman, "Financial Ratios, Discriminant Analysis and the Prediction of Corporate Bankruptcy," *The Journal of Finance*, Vol. 23, No. 4, 1968, pp. 589–609.
[64] Edward I. Altman and Edith Hotchkiss, *Corporate Financial Distress and Bankruptcy: Predict and Avoid Bankruptcy, Analyze and Invest in Distressed Debt*, Hoboken, N.J.: Wiley, 2006.
[65] Altman and Hotchkiss, 2006.
[66] Langabeer, 2006.
[67] Z. H. Ramamonjiarivelo, "Is Privatization the Solution to the Financial Distress of Public Hospitals?" *Academy of Management Proceedings*, Vol. 2014, No. 1, 2014.
[68] Langabeer et al., 2018.

$$Z = 6.56x_1 + 3.26x_2 + 6.72x_3 + 1.05x_4. \quad (6.1)$$

In Equation 6.1, x_1 is net working capital as a proportion of total assets ([current assets – current liabilities]/total assets), an important liquidity ratio. The variable x_2 is net assets as a proportion of total assets ([total assets – total liabilities]/total assets), a measure of financial leverage. The variable x_3 is earnings before taxes as a proportion of total assets ([total revenue – total expenses before taxes]/total assets), a measure of profitability and asset productivity. Tax expenses are not considered to improve comparison between for-profit and not-for-profit or public hospitals. The variable x_4 is the book value of equity as a proportion of total liabilities (book value of equity/total liabilities).[69]

Altman originally established Z-score thresholds to categorize levels of corporate financial distress leading to bankruptcy within two years.[70] Thresholds have since been updated based on subsequent data analysis from more-recent financial periods to reduce false-positive and false-negative prediction errors.[71] The thresholds are generally used to categorize Z scores into the following three categories: (1) no indication of financial distress, (2) potential for financial distress, and (3) severe financial distress.[72] Langabeer et al. uses 3.0 and 1.8 as Z thresholds for these three buckets, as applied to acute care hospitals; Z scores below 1.8 are categorized as severe financial distress. We use the same cutoffs in this analysis.[73]

Using this methodology provides a way to characterize the financial health of the hospitals, but there are limitations and caveats. Both Altman and Hotchkiss and Langabeer et al. recognized that it is better to tailor the Z-score model to specific industries to improve its accuracy and validity.[74] For our purposes, this would involve tailoring and validating the model to the California hospital industry in particular or the hospital industry at large, as researchers have recently done for rural hospitals.[75] Such an effort would require historical data of hospital financial health, bankruptcies, and closures. Given that we do not have historical bankruptcy or closure data and only use this model to determine general trends of financial distress and not to predict distress for individual hospitals, we do not alter or tailor this model and instead apply the Altman and Hotchkiss Z score exactly as Langabeer et al. do for Texas hospitals.[76] Although the application of the Z score to the California hospitals produces a higher proportion of hospitals with financial distress than the Langabeer et al. study found for Texas hospitals, we are unable to assess whether this difference is legitimate or the result of untailored models to specific state-

---

[69] Altman and Hotchkiss, 2006.

[70] Altman, 1968.

[71] Altman and Hotchkiss, 2006.

[72] The category names are adapted from Langabeer et al., 2018.

[73] Langabeer et al., 2018.

[74] Altman and Hotchkiss, 2006; Langabeer et al., 2018.

[75] G. M. Holmes, B. G. Kaufman, and G. H. Pink, "Predicting Financial Distress and Closure in Rural Hospitals," *The Journal of Rural Health*, Vol. 33, No. 3, 2017, pp. 239–249.

[76] Altman and Hotchkiss, 2006; Langabeer et al., 2018.

level hospital markets, given the lack of available historical bankruptcy and closure data. However, we show later in this chapter that the Z score aligns well with other financial metrics that measure elements of financial health, an indication that Z-score application to California hospitals captures most indicators of financial distress.

Additionally, we recognize that hospitals exhibit financial behavior that is different from the sample of corporations Altman has used to establish distress thresholds. Far fewer hospitals in severe financial distress are likely to close or go bankrupt than would be predicted for corporations in similar financial health, according to Altman.[77] A study of distressed community hospitals during the period 1983–1985 noted that 91 percent of these hospitals survived through the end of 1990.[78] More recently, Holmes, Kaufman, and Pink found that, among the 24 percent of rural hospitals identified as having medium or high financial distress from 2003 to 2013, only 3.4 percent actually closed.[79] Conversely, Altman's prior studies suggest that rates of bankruptcy within two years of financial distress are in the range of 70 to 80 percent.[80] Langabeer suggests that there are several reasons why formal bankruptcies do not occur as often for hospitals as with other companies.[81] Hospitals are organizations with complex business models substantially different from normal corporations and have strong ties to other community institutions, such as universities, churches, and nonprofit groups. Not-for-profit hospitals may benefit from philanthropic funds from other community organizations or private donors, and public hospitals can tap into tax revenue and other sources of public funds. Additionally, hospitals have a public health imperative to remain open more than other types of corporations and have some flexibility to adjust their financial behavior accordingly to fulfill this mission. For these reasons, hospitals do not often go bankrupt or close, but when they do, it can be particularly newsworthy and painful for communities.[82]

As previously stated, hospital financial statistics have limitations for objectively assessing the level of financial distress for a single hospital and do not allow for characterizing and comparing financial distress across a large group of hospitals. The advantage of the Altman Z score over other financial distress measures is that the Z score provides a general characterization of financial health that can be used to compare groups of hospitals. Despite the potential model-validity issues mentioned, the fact that the Z score is closely correlated with key financial and operational characteristics is an indication of its accuracy and usefulness.

The Z score can effectively link performance indicators to the spectrum of financial distress. Table 6.1 shows that the Z score is aligned with some of the leading financial indicators and

---

[77] Altman, 1968.

[78] G. Bazzoli and S. Andes, "Consequences of Hospital Financial Distress," *Hospital and Health Services Administration*, Vol. 40, No. 4, 1995, pp. 472–496.

[79] Holmes, Kaufman, and Pink, 2017.

[80] See, e.g., Altman and Hotchkiss, 2006.

[81] Langabeer, 2006.

[82] Langabeer, 2006.

ratios used to assess financial performance according to the Healthcare Financial Management Association.[83] These include the following:

- The *operating margin* measures the profitability of patient care operations and is often used to characterize financial trends within the health care industry. As expected, Z scores are closely correlated with margins across all hospitals.
- The *current ratio* (total current assets/total current liabilities) is an indication of liquidity and the ability to pay short-term obligations; a higher number indicates greater liquidity.
- *Debt-to-asset ratio* (total long-term debt/total assets) measures the extent to which a hospital is financially leveraged.
- The *average age of the plant* (accumulated depreciation/current-year depreciation expense) indicates the financial age of fixed assets and the extent to which an organization will need capital in the short term to invest in aging assets.
- *Days cash on hand* indicates the strength of financial reserves relative to operating costs, another measure of short-term liquidity. *Days in accounts receivables* is a measure of the efficiency of collections.
- The *average payment period* is a measure of how efficiently an organization pays its bills as it incurs costs, which is also partially a function of how efficiently it collects payment from accounts receivable.

For each of these measures, there is a strong correlation between the Z score and the performance area of the measure. There is obvious endogeneity in that correlation, given that the inputs for operating margin and current ratio are also in the Z-score calculation, but the correlation remains for the other performance statistics, including additional measures not listed here.

Table 6.1. Altman Z Score Alignment with Other Financial Health Indicators

Average Z Score in Past 5 Years	Operating Margin	Current Ratio	Debt-to-Asset Ratio	Average Age of Plant (years)	Cash on Hand (days)	Average Payment Period (days)
10>	11.7%	9.7	0.016	11.4	62	42
6 to 10	6.2%	3.8	0.126	11.1	74	65
2 to 6	2.4%	2.6	0.314	11.0	48	97
–2 to 2	–4.5%	1.3	0.468	11.9	23	161
<–2	–28.3%	0.7	0.432	14.3	12	331

NOTE: The value for each financial indicator for each Z-score group represents the hospital average indicator, averaged over the past five fiscal years.

[83] See Healthcare Financial Management Association, "Key Hospital Financial Statistics and Ratio Medians: Glossary of Formulas," webpage, undated.

To calculate Z scores for each hospital and other financial performance trends, we utilized publicly available data from all California general acute care hospitals reported to OSHPD via Hospital Annual Financial Disclosure Reports.[84] We used data from the five most recently reported fiscal years—2016–2017 is the most recently reported year and assumed to represent current fiscal status. Z-score input ratios were constructed using the following data fields from the OSHPD database: total current assets, total current liabilities, total assets, total liabilities, and net income before taxes and extraordinary items. Because of special reporting rules, all 36 Kaiser Permanente hospitals are grouped into Southern California and Northern California fiscal entities; we applied the same Z scores to all these hospitals and therefore did not pick up financial distress variations within the two regions. Financial data for all other MHSs were reported at the individual hospital level. We included 47 hospitals in the overall count that have been financially consolidated under a parent hospital but remain distinct facilities. For these hospitals, we reported the financial information from the parent facility. The resulting financial data can be mapped to 406 of the 418 general acute care hospitals for which OSHPD provides building-level data.[85]

Current Status of California Hospital Finances

California hospitals exhibit a wide degree of financial health. Figure 6.1 displays the Z score for the 2016–2017 fiscal year for all hospitals. Results are demarcated with different colors by their distress category, as defined by the Z-score thresholds. Figure 6.1 indicates that 66 percent of hospitals have no sign of financial distress, 12 percent have the potential for financial distress, and 22 percent have severe financial distress. These financial-distress percentages have remained generally stable, with slight reductions in distress over the past five years; the percentage of hospitals with the potential for or severe financial distress has decreased by 0.66 percentage points annually, on average, over the past five fiscal years.

[84] OSHPD, "Hospital Financials," webpage, undated.

[85] It is unclear why financial data for 12 hospitals were not available from the OSHPD data.

Figure 6.1. 2017 Financial Distress, by Hospital Category

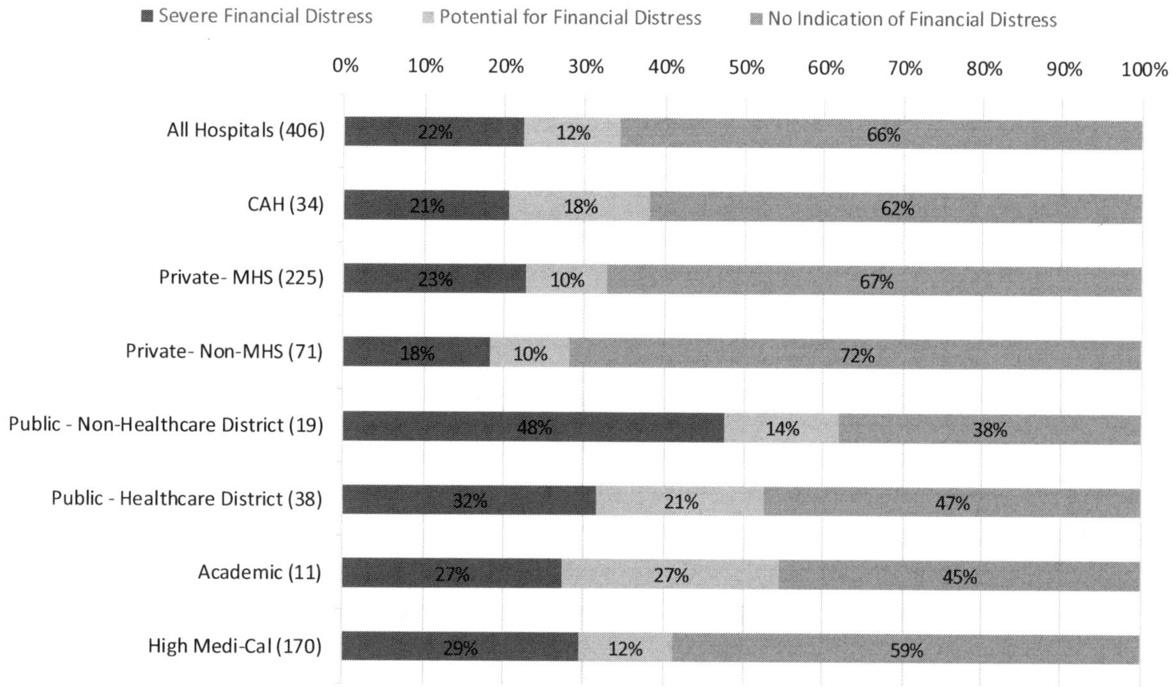

NOTES: We do not differentiate between private for-profit hospitals and private not-for-profit hospitals, because this is less relevant to interpreting the impact of financial distress on affordability than whether a hospital is independent or part of an MHS. Percentages are rounded and might not sum to 100. Hospitals can belong to more than one category.

Financial distress is unevenly distributed across hospital ownership models and types; this unevenness can disproportionately affect certain types of patient populations and communities. Compared with all California hospitals, a larger proportion of CAHs and public, academic, and high–Medi-Cal patient-day hospitals have the potential for financial distress. Note that these categories are not mutually exclusive, as 56 percent of CAHs are under public health care district ownership and most have high shares of Medi-Cal patients. Public hospitals have high Medi-Cal patient shares, which is aligned with their mission of providing safety-net services to local communities. Public hospitals averaged operating margin losses of 8.7 percent over the past five fiscal years, putting public hospitals in generally worse financial distress than private hospitals. Hospitals that serve more-vulnerable populations, such as CAHs and public and private high-Medi-Cal hospitals tend to have higher levels of financial distress. This can be partially attributed to lower reimbursement rates by Medi-Cal than from private insurers and higher rates of uninsured service provision than at private hospitals.

Individual hospital financial conditions alone do not define affordability. Financial health must be interpreted within the context of a hospital's ownership structure, which may influence its profitability, ability to secure funds, and ability to borrow capital for infrastructure projects. Among private hospitals that are part of an MHS, the financial condition of a single hospital within the system is less significant than the financial condition of the system as a whole. Systems may shift services and investments among hospitals that affect individual hospital profitability but optimize profitability for the system as a whole. The overarching financial structure also provides additional flexibility to diffuse financial distress across the system and raise or borrow capital that can be used for infrastructure projects at individual hospitals. This point is reflected in hospital credit rating data; approximately half of private MHS hospitals with some or severe financial distress retain investment-grade credit ratings as a result of being part of MHSs with good financial health. Despite operating some hospitals at little profit or a sustained loss, a system will take into account other such factors as referral networks and regional competitiveness when making decisions about individual hospitals that are struggling financially.

Alternatively, independent private hospitals lack the flexibility of MHS hospitals and entirely own the financial responsibility for seismic retrofit construction. Sixty-four percent of private non-MHS hospitals do not have credit ratings, and less than a quarter that have some or severe financial distress have investment-grade credit ratings. This means that seismic construction costs will likely impose more financial distress and affordability concerns on these hospitals than other private MHS or public hospitals with similar financial health. Independent private hospitals cannot handle negative operating margins like a private MHS hospital or some public hospitals, so financial distress imposes a tougher decisionmaking process.

The picture of affordability in the context of ownership structure is perhaps more bifurcated among public hospitals. Public hospitals have the potential for greater financial flexibility than private hospitals because they are backed by public entities with additional sources of revenue outside the hospital business enterprise, and they have the ability to raise revenue bonds or general obligation bonds if voters approve. However, among the 59 public hospitals in California, 38 (64 percent) are owned by hospital districts that often effectively lack these public-sector benefits. Hospital districts are special districts unique to California that do not have other sources of revenue and, more important, struggle from a lack of visibility to convince voters to approve bond-related tax increases for seismic retrofitting or other capital needs.[86] Health care district hospitals are often smaller and serve rural communities or populations on the fringe of metropolitan areas; half are designated as CAH hospitals. Despite their public-sector backing, 55 percent lack investment-grade credit ratings, and 40 percent do not have a credit rating at all. Twelve health care districts filed for bankruptcy from 1996 to 2014, some of which do not operate hospitals, and a few have recently dissolved.[87] Because of their low-income, high Medi-Cal-utilization populations, and significant constraints on raising other sources of revenue to finance large infrastructure projects, health care district hospitals likely have greater affordability concerns than other public or private hospitals with similar financial distress.

---

[86] This information is from anecdotal sources and hospital interviews. At least four health care district bond measures to raise capital for new buildings have failed over the past two election cycles. Bond measures must pass a two-thirds vote.

[87] Marc Joffe, "California Healthcare Districts in Crisis," California Policy Center, January 22, 2015.

Conversely, the remaining public hospitals owned by city or county governments or the University of California system are mostly large hospitals serving urban populations. They have much less difficulty raising capital via bonds, which in some cases might not be subject to ballot initiatives and might have the ability to access general-fund tax revenue. All have investment-grade credit ratings because they are owned by governments with strong fiscal health. Although 62 percent of these public non–health care district hospitals have the potential for or severe financial distress, their ability to raise additional sources of revenue significantly mitigates affordability concerns. Table 6.2 includes additional credit rating data related to each type of hospital discussed and summarizes general barriers and contributing factors to affordability that are explained above.

Table 6.2. Financial Distress in the Context of Hospital Ownership Type

Type of Hospital	% with Credit Rating	% with Investment-Grade Credit Rating	% in Some or Severe Distress with Investment-Grade Credit Rating	General Barriers and Contributors to Affordability
Private—independent	36	31	23	Barrier: Lack of financial flexibility Contributor: Can join hospital system to improve financial flexibility
Private—MHS	85	57	51	Contributor: Financial flexibility within system
Public—health care district	60	45	54	Barriers: Raising capital is politically difficult; small size
Public—non–health care district	100	100	100	Contributors: Owned by strong fiscal institutions; access to other public funds outside hospital revenue

Affordability of Seismic Requirements

The seismic retrofit requirements and financial distress disproportionately affect some hospital types, populations, and locations (see Table 6.3). For example, although 84 percent of all hospitals have some degree of SPC- or NPC-noncompliant buildings, the requirements disproportionately affect CAHs (91 percent), public health care district hospitals (97 percent), and high–Medi-Cal patient-day hospitals (92 percent). Twenty-eight percent of all hospitals require seismic action and either have the potential for financial distress or already have severe financial distress. Critical access, public, academic, and high-Medi-Cal hospitals are disproportionally affected by both seismic requirements and financial distress.

Table 6.3. The Distribution of Seismic Requirements and Financial Distress Across Hospital Types

Hospital Types (# of Hospitals)	% of Hospitals that: Have Severe or the Potential For Financial Distress	Are SB1953 Noncompliant	Are SB1953 Noncompliant and Have Severe or the Potential For Financial Distress	Are SB1953 Noncompliant and Have Severe Financial Distress
All Hospitals (406)	34%	84%	28%	18%
CAH (34)	38%	91%	35%	18%
Private- MHS (225)	33%	83%	26%	18%
Private- Non-MHS (71)	28%	87%	23%	14%
Public - Non-Healthcare District (19)	62%	89%	58%	47%
Public - Healthcare District (38)	53%	97%	53%	32%
Academic (11)	55%	82%	36%	9%
High Medi-Cal (170)	41%	92%	38%	28%

Displaying the concentration of seismic requirements and financial distress in Table 6.3 characterizes the distribution across ownership and hospital types. However, when characterizing the effects on the populations they serve, it is better to display results as the number and proportion of beds. The number of beds is a proxy for the size of the population served by a single hospital or a group of hospitals in a region and can be useful when determining how the combined effects of SB 1953 compliance and hospital financial distress is distributed among the population.

Table 6.4 shows that there is considerable variation of the effects of SB 1953 compliance and financial distress across populations and counties. The table groups California counties by size categories (≤150 beds, 151–800 beds, and >800 beds), roughly dividing the California counties into thirds. The table also shows the number of hospitals and the number of beds in each county for which we had OSHPD cost utilization data.[88] The table also displays

- the percentage of beds associated with noncompliant hospitals
- the percentage of beds associated with noncompliant hospitals that either have the potential for financial distress or already have severe financial distress
- the percentage of Medi-Cal patient-days associated with noncompliant hospitals that either have the potential for financial distress or already have severe financial distress.

These measures show how the effects of SB 1953 compliance and financial distress are distributed among counties and among Medi-Cal populations within counties. Ninety percent of all beds are associated with noncompliant hospitals, and most counties have well over 50 percent of hospital beds that are associated with noncompliant hospitals. Thirty-one percent of California beds are associated with noncompliant hospitals that either have the potential for financial distress or already have severe financial distress. This is unevenly distributed among the counties—some counties have none and others have 100 percent. Overall, 38 percent of Medi-Cal patient-days are associated with noncompliant hospitals with the potential for financial

[88] The OSHPD cost utilization data did not provide the number of beds for 58 of the 418 hospitals; we omitted these hospitals from the bed count but included them in the hospital count.

distress or current severe financial distress. Of the counties with more than 800 beds, Alameda, San Francisco, Stanislaus, and Tulare counties have at least 50 percent of their beds and Medi-Cal patient-days associated with noncompliant hospitals that either have the potential for financial distress or already have severe financial distress.

Table 6.4. Seismic Noncompliance and Financial Distress Distribution Across Counties

# of County Beds Category	County	# of Hospitals	# of Beds	% of Beds Associated with Non-compliant Hospitals	% of Beds Associated with Non-compliant Hospitals That Have Either the Potential for or Severe Financial Distress	% of Medi-Cal Patient-Days Associated with Non-compliant Hospitals with the Potential for or Severe Financial Distress
	Amador	1	52	0%	0%	0%
	Calaveras	1	48	100%	0%	0%
	Colusa	1	No Data	No Data	No Data	No Data
	Del Norte	1	42	0%	0%	0%
	Glenn	1	14	100%	0%	0%
	Inyo	2	62	100%	100%	100%
	Lake	2	50	100%	100%	100%
	Lassen	1	38	0%	0%	0%
<= 150	Mariposa	1	34	100%	100%	100%
	Mendocino	3	118	79%	21%	15%
	Modoc	2	113	100%	23%	35%
	Mono	1	17	100%	0%	0%
	Plumas	3	126	100%	0%	0%
	Siskiyou	2	61	54%	0%	0%
	Sutter	1	14	0%	0%	0%
	Tehama	1	76	100%	0%	0%
	Trinity	1	50	100%	0%	0%
	Butte	6	576	100%	52%	35%
	El Dorado	2	224	100%	50%	22%
	Humboldt	5	275	100%	28%	21%
	Imperial	2	268	100%	0%	0%
	Kings	1	230	0%	0%	0%
	Madera	2	464	23%	0%	0%
	Marin	4	458	90%	13%	29%
	Merced	2	230	19%	0%	0%
	Monterey	4	793	100%	12%	34%
	Napa	2	352	100%	41%	54%
151 - 800	Nevada	2	166	100%	37%	78%
	Placer	3	729	100%	9%	8%
	San Benito	1	181	100%	100%	100%
	San Luis Obispo	4	397	100%	28%	16%
	Santa Cruz	3	359	70%	0%	0%
	Shasta	5	712	100%	1%	0%
	Solano	5	672	79%	0%	0%
	Sonoma	8	769	67%	25%	17%
	Tuolumne	2	152	0%	0%	0%
	Yolo	2	156	100%	69%	0%
	Yuba	1	219	100%	0%	0%
	Alameda	17	3,424	76%	50%	62%
	Contra Costa	9	1,605	83%	24%	30%
	Fresno	10	1,983	99%	9%	15%
	Kern	13	1,769	97%	24%	41%
	Los Angeles	96	24,386	91%	30%	32%
	Orange	34	6,207	93%	37%	41%
	Riverside	20	3,722	93%	44%	49%
	Sacramento	10	3,016	98%	28%	46%
> 800	San Bernardino	24	4,081	82%	27%	35%
	San Diego	24	6,851	100%	26%	25%
	San Francisco	15	3,686	100%	74%	94%
	San Joaquin	7	993	100%	19%	17%
	San Mateo	7	1,781	74%	27%	31%
	Santa Barbara	6	1,071	79%	0%	0%
	Santa Clara	13	3,701	89%	17%	18%
	Stanislaus	8	1,416	95%	55%	56%
	Tulare	4	852	100%	68%	65%
	Ventura	10	1,573	83%	42%	63%
Total		418	81,414	90%	31%	38%

Framing Financial Distress and Affordability Changes Imposed by SB 1953

The affordability findings presented above are based on current hospital financial conditions before seismic compliance work for the 2030 deadline has been performed by hospitals. The expense of compliance work imposed on hospitals under the status quo policy will likely increase overall levels of hospital financial distress in California. It is impossible to predict what the actual financial landscape would look like postcompliance, but we attempt to provide a frame of reference to aid decisionmakers and indicate the types of hospitals that the seismic financial burden affects the most. The Z-score method incorporates financial ratios based on several key balance sheet and income statement terms, including total liabilities. Because SB 1953 compliance costs are a mandated future cost that hospitals must incur, we posit that this should theoretically be considered a long-term accounting liability,[89] and we extend the Z-score calculation to include estimated retrofit costs as a long-term liability for each hospital presented.

Using the estimated cost of compliance provides an estimate for the financial distress imposed on an individual hospital, all other things equal within its business model. In practice, hospitals can be adaptive to a certain extent, and we acknowledge it is highly unlikely that financially distressed hospitals would simply incur the liability without making financial or operational changes to offset the distress. When faced with an infrastructure cost that poses a significant financial burden, hospitals would be forced to financially adapt by reducing operating costs, cutting loss-leading services, reducing Medi-Cal services, forgoing other capital investments, merging with other hospitals, and so on, such that their balance sheets would not ultimately reflect the distress imposed by this estimation. However, the costs used to frame the increase in financial distress notably do not include escalation and financing costs, because of the uncertainty of these parameters. Including these additional costs would significantly increase the overall cost impact and long-term liabilities on hospitals' balance sheets, as discussed in the prior chapter. We expect this would further exacerbate the impact on certain types of hospitals, noted below.

Our cost estimates in the previous chapter indicate that cost is not equally distributed across hospital types, and we estimate that some hospital types receive more cost burden when normalized by hospital size. The cost burden statewide is $25 billion to retrofit all noncompliant buildings. Hospitals of different ownership types (e.g., MHS versus non-MHS, public versus private, public health care districts) have different opportunities and challenges when securing capital for construction projects. MHS and public hospitals can have access to funding outside their immediate hospitals and could potentially reduce the financial burden imposed by retrofit requirements. In addition, hospitals may choose to reduce capacity, given the imposed cost burden, which could disproportionally affect low-income or rural patients. The cost burden,

---

[89] Prior to retrofit construction and planning, the SB 1953 seismic requirements are a future financial liability. As construction progresses, the liability caused by the seismic requirements will slowly transition from a liability to an asset, depending on how the project is financed. However, for the purpose of estimating financial distress, we treat the seismic cost as a complete liability because we desire to assess the financial health of the hospitals currently and prior to construction start.

ownership structures, and patient-population types are all components of a hospital's ability to afford the construction.

When normalized by hospital size, the estimated retrofit cost is significantly more expensive for CAHs and public health care district hospitals. Table 6.5 displays hospital categories, estimated total retrofit cost in 2019 dollars (unescalated), total number of beds in the hospitals, and the average retrofit cost per bed. Note that, although Tables 5.2–5.6 refer to beds only in buildings requiring seismic upgrading, Table 6.5 shows the total number of beds for all hospitals in a given category. Normalizing the costs by the number of beds allows for a fair comparison across the hospital categories that sometimes have hospitals of very different sizes. On average, the estimated retrofit cost per hospital bed is $311,000. The CAHs' and public health care district hospitals' cost per bed is $647,000 and $481,000, respectively, because higher proportions of their hospital buildings require seismic retrofit.

Table 6.5. Estimated Retrofit Costs, by Hospital Category, Unescalated Costs

Category	Estimated Total Retrofit Cost	Total # of Beds	Estimated Average Retrofit Cost per Bed
All hospitals	$25.3B	81,414	$311K
CAH	$0.9B	1,467	$647K
Private, MHS	$16.0B	50,280	$318K
Private, non-MHS	$4.9B	14,624	$336K
Public, non–health care district	$0.9B	6,993	$130K
Public, health care district	$2.5B	5,145	$481K
Academic	$1.0B	4,372	$229K
High Medi-Cal	$10.5B	34,837	$300K

To modify the Z score with retrofit costs and estimate imposed financial distress, we added the estimated retrofit unescalated cost to total liabilities, which affects the variables x_2 and x_4 in the Z-score model. Results are presented in Figure 6.2, which incorporates current financial distress results presented in Figure 6.1 for comparison. Across all hospitals, the share of hospitals in severe financial distress increases by 18 percentage points, from 22 percent to 40 percent. The share of hospitals with no financial distress decreases by 18 percentage points, from 66 percent to 48 percent. All types of hospitals experience increases in financial burden, but changes are disproportionately borne by CAHs, private non-MHS hospitals, and public health care district hospitals. As previously stated, these hospital ownership types also are comparatively less capable of absorbing financial distress, which magnifies the impact of the increase in distress on overall affordability. This presents an obvious inequitable impact of the effects of SB 1953 compliance on certain geographic areas and lower socioeconomic populations served by these types of hospitals.

Figure 6.2. Financial Distress Estimates Imposed by SB 1953

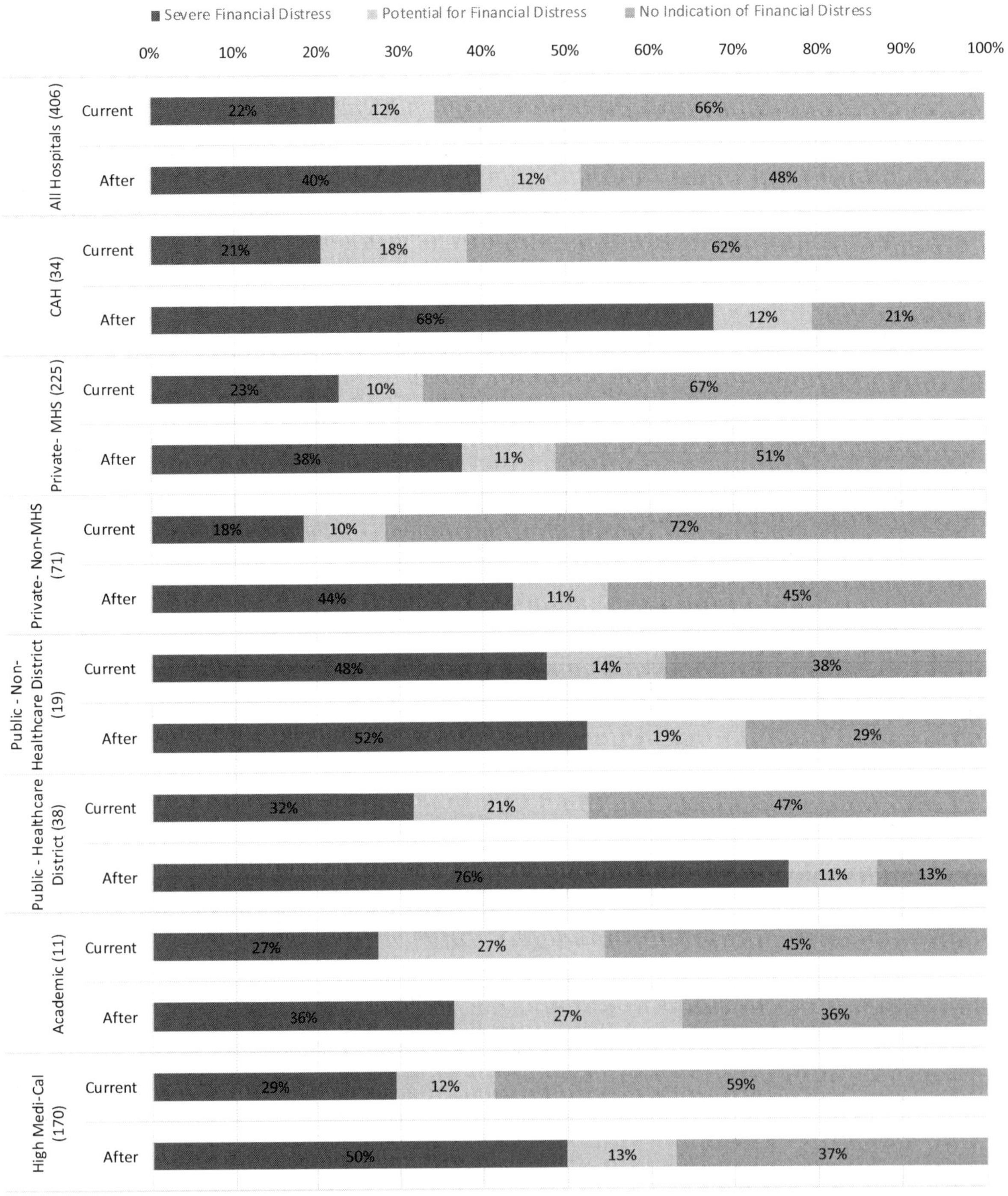

NOTES: Current = financial condition through fiscal year 2017. After = financial condition imposed by unescalated SB 1953 compliance costs, assuming 100 percent retrofitting. Percentages are rounded and might not sum to 100.

Conclusions on SB 1953 Affordability

This chapter presents analysis and findings that characterize the affordability of SB 1953 compliance across hospitals of different ownership structures and with different patient populations. The complex nature of affordability, and of the resulting hospital decisions and actions, make it difficult to accurately forecast the financial distress imposed by SB 1953 *for individual hospitals*. However, we characterized trends of financial distress across hospital types using published financial distress assessment methods that have recently been applied to acute care hospitals. The affordability research we presented shows that current and SB 1953–imposed financial distress have a tendency to affect certain types of hospitals and patient populations more than others.

Specifically, the affordability research has the following findings:

- **California hospitals exhibit a wide range of financial conditions, ownership types, and propensity for affording SB 1953 compliance.** Affordability and lack thereof are distributed unevenly across hospital attributes, patient populations, and locations.
- **Seismic compliance requirements seem affordable for approximately 48 percent of the California hospitals.** These hospitals either exhibit strong financial health, do not have major remaining compliance costs relative to their size, or both.
- **The SB 1953 2030 financial burden could yield as many as 40 percent of all hospitals in severe financial distress, up from 22 percent currently, an increase of 18 percentage points because of seismic requirement burden.** For these hospitals, there is a strong indication of unaffordability. Unaffordability will be more acutely experienced by financially distressed hospitals under certain ownership types, including independent private hospitals and public health care district hospitals. Conversely, SB 1953 will decrease the percentages of hospitals with no indication of financial distress from approximately 66 to 48 percent.
- **SB 1953 requirements disproportionally affect critical access, independent private, public health care district, and high-Medi-Cal hospitals.** If these hospitals respond to the increased cost burden by reducing hospital capacity, this will disproportionally affect rural and low-income patients who receive services from these types of hospitals, which presents an obvious equity concern.
- **SB 1953 disproportionally affects some geographic locations more than others.** We see considerable variation of noncompliance and financial distress across counties. Counties with large populations appear to have relatively more beds and higher percentages of low-income populations served by noncompliant hospitals in financial distress.
- **Because of SB 1953 requirements, financially distressed hospitals could take a number of managerial actions to improve financial health, some of which may adversely affect patient populations.** Some managerial actions could indirectly improve service provision by making hospitals more financially resilient, some would have no impact on service provision, and others would have an adverse impact on patient populations by (1) reducing hospital capacity by removing general acute care services from noncompliant buildings, (2) reducing or closing less profitable service lines, and (3) closing the hospital altogether. Additionally, deferment of capital investments in new buildings or technologies would indirectly present an opportunity cost to patients who

otherwise could have benefited from such investments. It is impossible for us to predict whether hospitals will take actions that have adverse impacts on patient populations or focus on other options to mitigate financial distress, but the unprecedented magnitude of the financial impact of SB 1953 on hospitals suggests that administrators will be forced to make difficult constrained decisions.

Affordability and knowledge of the disproportional effects of seismic requirement policies is important when considering the effects of SB 1953 and future seismic policies. Policies should balance seismic safety with potential unintended adverse effects on hospital capacity and patient populations. These results indicate that policies with more fidelity rather than unilateral seismic requirements can be mechanisms to achieve such a balance.

7. Learning from Approaches to Seismic Resilience in Other Sectors and Regions

A variety of reasons exist for the slow compliance behavior exhibited by hospitals, including the unfunded nature of the mandate, the extended time horizon of the 2030 deadline and extensions of previous deadlines, and the prolonged nature of complex hospital building projects. In assessing how to potentially improve compliance with SB 1953 in its current form or a future modified form, it is useful to look at other sectors of critical infrastructure that have also endured retrofitting to mitigate future earthquake damage and speed postearthquake recovery. This comparison helps further illuminate why hospitals have been so slow to comply with SB 1953 and why compliance will likely remain difficult under the status quo.

In addition to investments to retrofit hospitals, significant investments to retrofit infrastructure in California have been made among K–12 schools and higher-education institutions, state and local government buildings, mass-transit rail systems, roadway and rail bridges, dams and reservoirs, water conveyance systems, natural gas pipelines, and structurally deficient apartment and commercial buildings.[90] Despite widely varying types of assets, owners, and stakeholder groups involved in each infrastructure sector, the success of retrofit efforts in each sector seems to hinge on three general necessary conditions:[91]

1. comprehensive technical screening to assess seismic vulnerability and inform necessary structural modifications
2. an impetus to make necessary seismic retrofitting modifications
3. an ability to finance necessary structural retrofitting work.

Technical vulnerability screening has mostly been led by state agencies with regulatory jurisdiction over an infrastructure sector (e.g., OSHPD for hospitals), but in some cases local governments have performed this role, and in select cases individual actors (large water and energy utilities, for example) have performed their own vulnerability assessments. The impetus to make necessary modifications has involved a mix of legislative mandates to make modifications by future deadlines (similar to SB 1953) and voluntary actions spurred by regulatory bodies, stakeholder group interests, and internal management decisions.

---

[90] California Seismic Safety Commission, *Progress Report for the California Earthquake Loss Reduction Plan: 2002–2006*, Sacramento, December 2003.

[91] This framework is adapted from evidence in California infrastructure sectors described in this chapter. It is strengthened by speculation from Oregon policymakers in the state's 2013 Resilience Plan that a lack of progress in each of these factors has limited the state's infrastructure retrofitting efforts to reduce seismic risk (Oregon Seismic Safety Policy Advisory Commission, *The Oregon Resilience Plan: Reducing Risk and Improving Recovery for the Next Cascadia Earthquake and Tsunami*, Salem, February 2013).

Financing seismic retrofit projects is closely aligned with other forms of large-scale infrastructure investments, for which there are generally three theoretical principles of who should bear the costs:

1. User-pay principle: Users of the asset receive most or all of the benefits and thus directly pay for the full cost of the goods or services they consume (e.g., road tolls, train tickets, utility rates).
2. Indirect beneficiary pays principle: Actors indirectly benefiting from an asset contribute up to the marginal economic value derived from it (e.g., local taxes on land-value increases from new transportation development).
3. Public financing: Tax revenue is used to pay for infrastructure when benefits are widespread (pure public goods—e.g., military bases) or when collecting user or beneficiary fees is impossible or inefficient (e.g., interstate highways).[92]

Where private ownership of public-good infrastructure exists (as often occurs in the U.S. health care system), owners usually must rely solely on the user-pay principle or external sources of funding, such as investment or donations, unless the public sector determines that a certain level of public financing is appropriate.

The following sections detail progress made in seismic retrofitting in several different critical infrastructure sectors in California.

Dams

More than 1,200 dams in California fall under the jurisdiction of the Department of Water Resources Division Safety of Dams (DSOD) based on their size and volume of water. These include all dams owned and operated by the state and local governments and public and private water and energy utilities. The federal government also owns and operates numerous dams under the jurisdiction of the Federal Energy Regulatory Commission, Bureau of Reclamation, and Army Corps of Engineers. The vast majority of these structures were built 50 to 100 years ago, before engineering and construction practices evolved to mitigate seismic risk.

Dams pose a greater catastrophic seismic risk than any other type of infrastructure asset because of the potential for widespread loss of life, economic damage, and environmental damage in the event of a dam failure. The seismic vulnerability of dams has been a concern of state engineers since at least the 1920s, when the St. Francis dam collapsed in Southern California as a result of a nonseismic sudden failure attributable to poor construction.[93] A renewed interest in dam seismic vulnerability began in 1971, after the San Fernando earthquake nearly resulted in disastrous performance of multiple dams in Southern California. This

---

[92] Hirotaka Yamauchi, "Characteristics and Cost Sharing of Transport Infrastructure," *JBIC Review*, No. 3, December 2000, pp. 41–51.

[93] Donald H. Babbitt, *Improving Seismic Safety of Dams in California*, Sacramento: California Department of Water Resources, Division of Safety of Dams, 1993.

prompted the first mandated review of similarly designed dams for seismic vulnerability, which led to numerous seismic retrofits of dams and some dams being removed from service or replaced.[94] The 1989 Loma Prieta and 1994 Northridge earthquakes prompted a further review of dams located close to fault lines and an increase in inspections of high-risk dams in zones of high seismic hazard.[95] Over the past two decades, DSOD has conducted in-depth investigations and reevaluations of numerous dams, often requiring years of work and millions of dollars for larger structures. This work, along with DSOD's physical inspections of all dams under its jurisdiction, has resulted in more than $1.5 billion in seismic improvements to dams.[96] Federal operators have also continued to conduct dam inspections and have performed several seismic retrofit projects.[97]

Efforts to retrofit dams to reduce seismic vulnerabilities have generally been successful in that, where needs have been identified, they seem to have been met with an appropriate response within an appropriate amount of time. This success can be analyzed through the aforementioned three necessary conditions of retrofit programs. DSOD has conducted comprehensive reevaluations of seismically vulnerable dams in conjunction with its typical physical inspections of all dams under its jurisdiction to identify dams in need of structural retrofitting. Federal operators have conducted similar investigations for dams under their jurisdiction. The release of dam condition assessment ratings has effectively served as the impetus for state and federally owned dams to perform seismic retrofits where necessary. For local government or utility-owned dams, DSOD does not mandate retrofitting but instead exercises its power to force dam operators to lower water levels to acceptable risk levels, which serves as an impetus by effectively reducing the storage capacity of the dam. Such an action appears to have been a driving motivation for the San Francisco Public Utilities Commission to replace its Calaveras Dam, which DSOD had forced to reduce water levels to 40 percent of capacity since 2001.[98] Additionally, a dam-condition rating that indicates the presence of a seismic vulnerability may sufficiently spur local owners to conduct engineering evaluations and perform seismic retrofits where necessary. In Napa, in 2017, a less than satisfactory rating of one dam motivated utility managers to conduct further assessments of the dam.[99]

Finally, funding sources for dam retrofits appear to be plentiful where needs are identified. State and federally owned dam projects are easily publicly financed by general-fund tax revenue or bond measures through the budgets of jurisdictional agencies. State and federal owners appear to have financed all retrofit projects for which jurisdictional agencies have suggested there is a need. Locally owned dam retrofit projects can be funded by either public financing or the user-

---

94 Babbitt, 1993.

95 California Seismic Safety Commission, *Northridge Earthquake: Turning Loss to Gain*, Sacramento, 1995.

96 California Department of Water Resources, "Lake Perris and Perris Dam Projects," webpage, undated.

97 Alastair Bland, "How California Water Suppliers Are Getting Earthquake-Ready," KQED Science, April 18, 2018.

98 Dominic Fracassa, "$823 Million, 31 Billion-Gallon Calaveras Reservoir Dam Ready for Debut," *San Francisco Chronicle*, September 21, 2018.

99 Howard Yune, "California Rates Napa's Lake Hennessey Dam in 'Fair' Condition," *Napa Valley Register*, September 5, 2017.

pay principle via increased utility rates. Although dam capital projects are relatively expensive, utilities that own and operate these assets typically have customer bases commensurate with the size of the dam. Capital costs can be equitably dispersed among current and future customers via debt financing at affordable costs.

Water Conveyance Infrastructure

Closely associated with dams is the conveyance infrastructure network of pipelines, aqueducts, and pumping stations that moves water between dams and bodies of water to urban and agricultural users. These systems are vital to the California economy that has developed around major urban areas that lack adequate local water supply. The main seismic vulnerabilities within these systems are the multiple points at which critical conveyance networks cross fault lines on the path to transporting water to the Bay Area and Southern California metropolitan regions. For the Bay Area, major pipelines carrying water from the Sierra Nevada mountains cross East Bay fault zones on their way to providing water to 4 million Bay Area residents, and multiple reservoirs sit directly on top of fault line depressions.[100] For Southern California, the State Water Project aqueduct, Los Angeles Aqueduct, and Colorado River Aqueduct each cross the San Andreas fault at different points on the way to providing water to 19 million people. A major earthquake in either region could potentially sever these supply lines, cutting off water to the respective metropolitan regions for an extended period. Experts posit that attempts to seismically retrofit these infrastructure networks will, at best, mitigate the impact of a major earthquake on water supplies rather than wholly prevent damage and impacts, similar to efforts in other infrastructure sectors.[101]

Jurisdiction over these infrastructure assets is primarily controlled by regional wholesale water providers and utilities: the San Francisco Public Utility Commission and the East Bay Municipal Utility District (in the Bay Area) and the Metropolitan Water District and the Los Angeles Department of Water and Power (in Southern California). Interestingly, progress on seismic retrofitting of vulnerable conveyance infrastructure has so far differed significantly between the two regions, with Southern California lagging behind the Bay Area.[102] The East Bay Municipal Utility District has spent $350 million building backup tunnels, flexible piping systems, and new aqueducts, and the San Francisco Public Utility Commission has made significant progress on a $4.8 billion plan, including a new tunnel across the San Francisco Bay, flexible piping systems within tunnels, and a completely replaced reservoir that had seismic concerns related to dams.[103] Comparatively, the Los Angeles Department of Water and Power

---

[100] Bland, 2018.

[101] Chris Austin, "Earthquake Resilience of Southern California's Water Distribution Systems," Maven's Notebook, January 3, 2018.

[102] Rosana Xia and Rong-Gong Lin II, "Earthquake Could Destroy L.A.'s Water Lifeline," *Los Angeles Times*, December 15, 2014.

[103] Bland, 2018.

and the Metropolitan Water District have so far engaged only in preliminary planning and project development to identify engineering solutions to seismic vulnerabilities in the conveyance systems.[104] However, the Metropolitan Water District has long been the main proponent backing the California Water Fix project, for which seismic vulnerability improvement is one of several goals.

In assessing the success of retrofitting Bay Area conveyance infrastructure and the delayed implementation of retrofitting Southern California conveyance infrastructure, it is again useful to assess the aforementioned necessary conditions. In this case, there are few to no capacity limitations for comprehensive vulnerability screening, given that each entity owns the responsibility for managing its infrastructure and has sufficient expertise (or access to expertise) to analyze its seismic vulnerability. Additionally, there are few to no limitations on funding, since each utility can spread costs among its large customer base. For instance, the Metropolitan Water District estimates that its 26 percent share of the $17 billion Water Fix project would cost its average household only an additional $3 dollars per month.[105]

The driving factor seems to be differing levels of impetus to make necessary seismic retrofitting modifications between the two regions. Following the 1989 Loma Prieta earthquake, the East Bay Municipal Utility District performed a vulnerability assessment and calculated a six-to-one benefit-to-cost ratio for implementing a (now completed) seismic improvement program to mitigate future earthquake losses. In the case of the wholesaler San Francisco Public Utility Commission, many of its customer agencies were worried about water reliability amid the seismic vulnerability of neglected infrastructure.[106] They effectively lobbied the California state legislature to pass AB 1823 in 2002, mandating a capital improvement program to improve the water storage and conveyance system that was "designed to outdated seismic safety standards" in consultation with the state Seismic Safety Commission to prevent disastrous public health and economic consequences.[107] The bill provided a compliance timeline requiring completion by 2015, a reasonable amount of time for such a large capital improvement plan. It is unclear why the legislature singled out the San Francisco Public Utility Commission without also addressing similar vulnerabilities within Southern California conveyance infrastructure.

Bridges

Seismic risk to transportation systems is magnified by the potential for lost lives and economic system disruption. The performance of highway bridges is a critical determinant of the

---

[104] Bland, 2018.

[105] Metropolitan Water District of Southern California, *Modernizing the System: California Waterfix Finance and Cost Allocation*, Los Angeles, 2017.

[106] California Planning and Development Report, "Cities Pressure San Francisco to Repair Hetch Hetchy," April 1, 2002.

[107] California Assembly Bill 1823, An Act to Add and Repeal Division 20.5 (Commencing with Section 73500) of the Water Code, Relating to Regional Water Systems, California General Assembly, January 18, 2002.

seismic performance of the entire transportation system. Because most highway bridges predated the evolution of national highway seismic standards developed in the early 1990s, numerous bridges in seismically active areas are vulnerable to collapse or major damage in an earthquake.[108] California has been more proactive than other states and the federal government, having adopted seismic design practices in the early 1970s following the 1971 San Fernando earthquake that kick-started bridge retrofitting. These efforts were significantly increased following notable bridge and overpass failures in the 1989 Loma Prieta earthquake (Bay Bridge collapse) and 1994 Northridge earthquake (I-10 and I-5 collapses).[109] Over the past 30 years, California has spent more than $12 billion (without accounting for inflation) on bridge seismic retrofit programs.[110] At nearly 90 percent complete, with remaining work fully funded, the programs seem to have been quite successful in achieving the goal of reducing bridge seismic vulnerability within a reasonable time frame.

Similar to dam and hospital screening efforts, the programs have been managed by the state jurisdictional agency, the California Department of Transportation (DOT), which screened all of the 12,000 state-owned bridges in the 1990s, as mandated by state legislation (SB 36X, 1989), and identified roughly 2,200 in need of being retrofitted to meet the latest seismic safety standards.[111] DOT additionally screened numerous local government-owned bridges and identified 1,200 more bridges in need of retrofitting. Bridge retrofits were split into three separate programs: state-owned bridges, toll bridges, and locally owned bridges—with different funding streams for each. The impetus for the programs was also strong. Significant economic damage and loss of life occurred from the Loma Prieta and Northridge earthquake bridge collapses, but officials also noted that bridges that had previously been retrofitted incurred little to no damage, proving the value of investing in the projects before damage occurs.[112]

Adequate funding was provided for the programs via several different sources, using user-pay and public-financing principles. By far the most expensive component of the programs is the $9.4 billion cost of retrofitting the state's nine major toll bridges, including $6.5 billion for replacing the Bay Bridge. A mix of toll user fees (64 percent), state and federal highway funds (28 percent), and proposition bonds (8 percent) was used to allocate the cost burden between individual users and the state and federal governments.[113] The state bridge program, costing $2.8 billion, was funded by a mix of proposition bonds (69 percent) and state and federal highway funds (31 percent). The local bridge program, costing $1.35 billion, was primarily covered by the

---

[108] William J. Mallett, Nicole T. Carter, and Peter Folger, *Earthquake Risk and U.S. Highway Infrastructure: Frequently Asked Questions*, Washington, D.C.: Congressional Research Service, May 2016.

[109] Virginia Ellis, "1,364 Additional Bridges Are Put on Retrofit List: Earthquakes; Caltrans Report Says 302 Highway Structures Needing Seismic Strengthening Are in L.A. County; Plans Call for an Accelerated Schedule That Would Be Completed by 1997," *Los Angeles Times*, July 1, 1994.

[110] DOT, "Seismic Retrofit Program," June 2014.

[111] DOT, 2014.

[112] Ellis, 1994.

[113] DOT, 2014.

federal Highway Bridge Program, with a requirement that local governments provide an 11.5 percent cost match. California voters passed Proposition 1B in 2006 to provide funds to cover this local matching requirement.[114] The mixture of funding sources used to pay for these retrofit programs demonstrates how an expensive infrastructure program can be financed following infrastructure financing principles that reasonably share the cost burden among individual users, multiple levels of government, and voter-approved initiatives.

Housing

Although housing is not a critical infrastructure asset or system, a major seismic vulnerability exists within a significant portion of California's building stock that was built before structural codes evolved in the 1980s. The most widespread type of building is a wood-frame "soft-story" structure, called this because of a critical deficiency in the strength of the first story.[115] These structures are commonly used as apartment buildings throughout metropolitan areas in California. Additional vulnerable building types include nonductile concrete, steel moment frame, and concrete tilt-up structures. Several hundred thousand apartment units are estimated to exist within these structures around the state.[116] In the 1994 Northridge earthquake, an estimated 49,000 apartment units in the region were destroyed or seriously damaged, roughly two-thirds of which were soft-story structures, including one that collapsed, killing 16 people.[117] Experts warn that a much larger earthquake could lead to a twofold vulnerability: an immediate loss of life and injuries from collapses and a long-term housing crisis, as significant portions of housing would be damaged beyond repair.[118] The latter impact would have widespread economic and social consequences similar to if many hospitals in a region had extremely reduced capacity or were rendered useless after a major seismic event.

Multiple past efforts within the legislature have failed to introduce statewide retrofit programs because of a concern of imposing significant costs on owners and tenants.[119] The movement to impose mandatory retrofitting has instead been led by a handful of cities, starting with San Francisco's program that began in 2013 and imposes retrofitting for 5,000 soft-story apartment buildings in the city. Similar programs have been initiated by Berkeley, Alameda, Oakland, and Fremont in the Bay Area and Los Angeles, Long Beach, Santa Monica, West Hollywood, Beverly Hills, and Burbank in the Los Angeles metropolitan region. Los Angeles's

---

[114] California Transportation Commission, "Adoption of Proposition 1B Local Bridge Seismic Retrofit Guidelines," Resolution LBS1B-G-0708-01, May 29, 2008.

[115] Federal Emergency Management Agency (FEMA), *Seismic Evaluation and Retrofit of Multi-Unit Wood-Frame Buildings with Weak First Stories*, Washington, D.C., May 2012.

[116] California Seismic Safety Commission, 2003.

[117] Mayoral Seismic Safety Task Force, *Resilience by Design*, Los Angeles, Calif., December 2014.

[118] Rong-Gong Lin II, "Big Earthquake Would Topple Countless Buildings, but Many Cities Ignore the Danger," *Los Angeles Times*, January 17, 2019.

[119] Lin, 2019.

program has mandated retrofits for more than 15,000 buildings. After passing an ordinance establishing the program, building officials in each city screen all relevant properties to determine which structures require retrofitting and establish future deadlines for classes of buildings. Property owners are required to have licensed engineers perform a structural analysis and produce construction documents for contractor construction. This process has potentially been aided by collaborative work between FEMA and Bay Area interest groups that produced a centralized repository of best practice retrofitting techniques for the San Francisco program.[120] The majority of retrofit projects are relatively simple and do not displace tenants during construction. Officials in Santa Monica have estimated that costs per apartment unit range from $5,000 to $10,000 for noncomplex, two-story structures.[121] Actual costs in San Francisco, however, have been as high as $50,000 per unit.[122] Cities have adopted different policies on passing on costs to renters, with some allowing it (San Francisco), others establishing cost sharing between owners and renters (Los Angeles), and others forcing owners to pay the entire cost (Santa Monica).[123]

It is too early to draw conclusions about the degree to which these programs will be successful in achieving retrofit compliance because the earliest program (San Francisco) has not reached its deadline, and the remaining programs have deadlines extending well into the future. However, results thus far in San Francisco indicate that its program is working well. As of January 2019, 53 percent of buildings have completed retrofit work, 40 percent have submitted paperwork to meet future deadlines, and only 6 percent are currently noncompliant.[124] Thus, it seems plausible that other cities can achieve similar results. These results again demonstrate the importance of effective technical screening and assistance from jurisdictional authorities driven by the impetus of a mandate. This presents an interesting example, with respect to funding, of private infrastructure owners being able to finance retrofit projects to meet an unfunded mandate when the projects are affordable. Costs for apartment building retrofits are not unduly expensive when considering the current value of the buildings and the extent to which real estate equity values have appreciated recently in California. Even when owners are required to pay for entire projects without passing on costs to renters, it seems plausible that they should be able to afford complying with the mandate.

---

120 FEMA, 2012.

121 Santa Monica Rent Control Board, "Study Session on Consideration of Tenant Protections and Costs Associated with the City of Santa Monica's Mandatory Seismic Retrofit Program," letter from Tracy Condon to the Rent Control Board commissioners, October 12, 2017.

122 J. K. Dineen, "For SF Landlords, Deadline Looms to Retrofit Soft-Story Buildings," *San Francisco Chronicle*, April 9, 2017.

123 Lin, 2019.

124 San Francisco Department of Building Inspection, "Soft Story Properties List," webpage, undated.

Conclusions for SB 1953 Compliance

Discussions surrounding SB 1953 compliance have often failed to acknowledge the tremendous amount of work performed in these other infrastructure sectors to reduce major seismic vulnerabilities that could have equally significant impacts on Californians. In some sectors, such as dams and highway bridges, requirements covering every asset across the industry are just as robust as efforts to protect every acute care hospital. In other areas, such as housing and water conveyance, there has been less of an effort to holistically protect every asset that may be vulnerable.

One clear conclusion is that progress in meeting seismic retrofitting goals in these other sectors has far exceeded progress among hospitals. Where state and local authorities have set out to significantly reduce seismic vulnerabilities in key infrastructure areas, they have generally been successful in all areas except hospitals. This begs the question of what is different about the hospital sector or elements of the hospital seismic compliance program that has led to this gap in progress.

Similar to bridges and dams, hospitals have been thoroughly screened and classified according to seismic vulnerability by a jurisdictional authority, so this is not an issue. Additionally, the impetus for performing retrofit work clearly exists via the SB 1953 mandate; hospitals know what level of compliance they need to achieve by fixed future deadlines. There may be some confusion about how to actually achieve compliance, given the complex nature of hospital building projects and the evolution of OSHPD ratings. This seems distinct from other sectors, and it may be contributing to delays in compliance. Yet the most obvious distinction between hospitals and other sectors is in the ability to finance the necessary retrofit projects.

First, before discussing individual project financing, a comment is warranted on overall sector costs. As has been noted within the hospital industry, total costs are quite large. Total spending on hospital retrofit projects will likely end up being orders of magnitude higher than spending on retrofits in other infrastructure areas. A 2003 Seismic Safety Commission report estimated that $19 billion had been spent from 1990 to 2002 on overall infrastructure retrofitting as the state ramped up its effort to mitigate asset vulnerability.[125] Hospital retrofitting was estimated (using the 2002 RAND estimate) to have made up 37 percent of this total, more than double spending on the second-largest category (bridges). In its future projections, the Seismic Safety Commission report estimated that between $21 billion and $35 billion would be spent, mostly driven by between $10 billion and $24 billion spent on hospitals (the range from the 2002 RAND report).[126]

As other major infrastructure retrofit programs have neared completion and updated estimates of SB 1953 compliance costs were developed for this study, it remains clear that these costs will greatly exceed spending on seismic retrofitting in any other sector and perhaps exceed

---

[125] California Seismic Safety Commission, 2003. All costs mentioned in this paragraph reference uninflated 2003 costs from the report.

[126] Meade, Kulick, and Hillestad, 2002.

all other sectors combined. While this is an interesting and noteworthy finding, given that other sectors are at least equally as important as critical infrastructure systems, it is not necessarily surprising. The complexity of hospital buildings drives up retrofit costs relative to other types of buildings, and SB 1953 nonstructural regulations (which this report estimates as roughly half of total costs) produce a type of regulatory burden that does not exist in other sectors.

The large overall sector cost in and of itself is not necessarily a problem. The ability to finance individual projects is distinct from overall spending—hospitals must simply finance the amount of their own projects, not finance all projects as a sector. Thus, the problem is more directly that hospitals are very constrained in their ability to pass on the costs of these infrastructure projects through their business models to the private actors that benefit from retrofitting. In other sectors, asset owners have been able to finance their retrofit projects by utilizing either user-pay or public-financing principles. In these situations, the private beneficiaries of such programs pay the costs.

Because of the nature of health care funding, hospitals are very constrained in their ability to pass an increase in costs to their customers using the user-pay principle. Government and private insurance rates are set through a complex policy and negotiation process that does not take into account increases in seismic compliance costs. Some hospitals may decide to implement managerial actions that generate higher profit margins to increase funding for infrastructure projects, but this may be insufficient. In regard to public financing, only a handful of hospitals across California owned by state, county, or municipal governments have the ability to access general-fund tax revenue for retrofit projects. The largest group of public hospitals, facilities owned by health care districts, have thus far struggled politically to convince voters to support public financing for retrofit projects.

Additionally, hospitals have little ability to pass on costs to indirect beneficiaries—e.g., current and future populations that would need to use health services after a major seismic event. Thus, most hospitals are forced to pay for expensive retrofitting projects by shifting services to generate additional revenue, using reserve funds, or accepting an increase in debt financing costs that results in lower operating margins. Given the strain that each of these options places on the hospital business model, it is not surprising that many have chosen to delay costly seismic compliance projects. This evidence may warrant a policy discussion of who stands to benefit from hospital seismic retrofitting and how those benefits might be captured to shoulder some of the project costs for each type of hospital.

8. Policy Alternatives for Seismic Resilience

The Alquist Act was motivated by the collapse of hospitals during the Sylmar earthquake of 1971. Later, the presence of nonstructural damage that compromised hospital operations in the wake of the 1994 Northridge earthquake prompted the adoption of SB 1953. According to OSHPD documentation notes, the law's goal is to "have general acute care hospital buildings that not only are capable of remaining intact after a seismic event, but also capable of continued operation and provision of acute care medical services after a seismic event."[127] The purpose of this chapter is to identify and briefly analyze some possible alternative approaches to reaching the goals set out by SB 1953.

Given this report's focus on cost and affordability issues, the options presented in this chapter aim to achieve the policy's goals while either reducing costs or redistributing costs across stakeholders. Before proceeding with that analysis, we note three big-picture policy observations we do not directly address in this report because of limitations in scope and timeline. These issues broadly frame the debate about SB 1953 and alternatives to it.

First, although hospitals provide specialized care services not elsewhere available within a community, they also depend on a variety of infrastructure, services, and other institutions. Hurricane Maria in Puerto Rico in 2017 and similar experiences show that interruptions in access to more-routine care for chronic conditions can significantly elevate morbidity and mortality risk for months following natural disasters.[128] Increasingly, care for such services is provided in nonhospital clinical settings. Policymakers may therefore wish to consider a more holistic approach to resilience and disaster preparedness that situates hospitals within this context and focus on the overall capacity of communities to continue to function in both the immediate aftermath and the longer term after an event.

Second, although detailed cost-benefit analysis of SB 1953 is outside the scope of this analysis, policymakers should consider whether the benefits of the additional upgrades required by the law before 2030 exceed the costs at a statewide level. Efforts made, to date, to bring hospitals into compliance with SB 1953 represent sunk costs. When considering whether to continue with the policy as is or to make changes, policymakers can take these efforts as given and focus only on further costs to be incurred versus the benefits they will yield.

Finally, even presuming that the net benefits of the policy are positive, it is worth considering whether investing in additional infrastructure upgrades constitutes the best use of limited resources. Capital set aside for hospital seismic compliance might have more public health

---

[127] OSHPD, "Seismic Compliance Program Overview," Sacramento, 2015.

[128] See, for example, Carlos Santos-Burgoa, Ann Goldman, Elizabeth Andrade, Nicole Barre, Uriyoan Colon-Ramos, Mark Edberg, Alejandra Garcia-Meza, Lynn Goldman, Amira Roess, John Sandberg, and Sco Zeger, *Ascertainment of the Estimated Excess Mortality from Hurricane Maria in Puerto Rico*, Washington, D.C.: George Washington University, Health Sciences Research Commons, 2018.

impact if invested in more-routine health-promotion activities. Policy has a clear role in prompting investments in seismic resilience as a public good, but hospitals' ability to make other investments that would improve health outcomes decreases as mandates grow more restrictive.

Framework for the Evaluation of Policy Alternatives

We identified policy options through a review of the academic and grey literature, including peer-reviewed journal articles, policies, and plans from the United States and (to a lesser extent) other countries with significant seismic risk. We also elicited ideas during the aforementioned interviews with hospital officials, academic experts, and other stakeholders. Given the scope and timeline of the project, we did not seek to generate a comprehensive list of alternatives. Instead, we present a short list of options that illustrate a range of policy instruments.[129]

As a benchmark for comparison, we begin by assessing the status quo. Next, we analyze four broad sets of alternative policy options that policymakers may adopt to achieve the goals of SB 1953 while addressing some of the cost and equity concerns outlined above:

1. public-financing options that spread the costs of compliance among a larger set of stakeholders
2. approaches that relax the requirement that every hospital achieve compliance in favor of a more regionalized approach that targets enforcement based on each facility's place in the larger system of local care
3. options that extend regulatory timelines, offering hospitals more flexibility over when to make the seismic retrofit investments
4. options that might speed up OSHPD's process for reviewing and approving seismic upgrade plans.

Beyond these policies, a full repeal of SB 1953 could also be considered. We do not address this option. Evaluating the likely costs and benefits of this option would require considering the likelihood and magnitude of earthquakes, likely short- and longer-term morbidity and mortality result from an earthquake, and other issues that were beyond the scope of this analysis.

We evaluate each set of policy options on the following set of standard policy analysis criteria:[130]

- **Effectiveness:** Qualitatively assess the likelihood that hospitals affected by the policy could continue operations immediately after the earthquake and during the months or years needed for full disaster recovery.

---

[129] Peter Bohm, and Clifford S. Russell, "Comparative Analysis of Policy Instruments," in Allen V. Kneese and James L. Sweeney, eds., *Handbook of Natural Resource and Energy Economics*, Amsterdam: Elsevier, 1985, pp. 395–460.

[130] Eugene Bardach and Eric M. Patashnik, *A Practical Guide for Policy Analysis: The Eightfold Path to More Effective Problem Solving*, Washington, D.C.: CQ Press, 2005.

- **Cost:** For the purposes of evaluating alternatives, focus on the main components of social costs. These are the direct costs of implementing the policy relative to the status quo and the likely change in the risk of hospital closures.
- **Distributional effects and equity:** Assess how the benefits and costs of the options are distributed among hospitals, regions, and the state.
- **Feasibility:** Qualitatively assess the likely administrative barriers, including informational intensity, monitoring and enforcement, and political considerations.
- **Other considerations:** Assess the likely effects on entities not directly targeted by the policy and other unintended consequences.

The Status Quo

The details of SB 1953 are described in detail earlier in the report. In summary, the seismic standards apply to every hospital providing inpatient general acute care. By 2030, hospital buildings that are not in substantial compliance with these standards (SPC-2 buildings) must be retrofitted (typically to SPC-4D or higher standards), demolished, replaced, or converted to nonacute use. Failing that, hospitals then face a notice of violation, and the California Department of Health Services is required to suspend or refuse to renew the license of a hospital that has received such a notice.

Effectiveness

Assuming that SB 1953 is fully enforced, it will eventually be effective in meeting its goal of 100 percent compliance among active hospitals. However, interviews conducted for this report suggest that some market participants expect that the policy may be modified to reduce the risk of hospital closures and provide hospitals some kind of flexibility in compliance. Because of prior deadline extensions, past experience reinforces these expectations and may lead some hospitals to delay their retrofit investments in hopes of relief. Policymakers may resolve the uncertainty in the market by increasing the credibility of enforcement or explicitly adopting another policy alternative.

Cost

The direct cost of the status quo policy in equilibrium is the number of seismically upgraded units supplied multiplied by the cost to upgrade per unit. Because some hospitals have already upgraded or were already built to meet SB 1953 standards, some portion of the policy's costs have already been realized and can be viewed as sunk by decisionmakers—that is, the costs that have already been spent and cannot be recouped and, thus, should not factor into decisions going forward. To provide the most-relevant information for considering policy alternatives, we estimate the remaining direct costs for the status quo. In this report, we find that if every hospital needing to upgrade complies with SB 1953, the total cost will range between $34 billion and $143 billion by 2030.

Attainment of 100 percent compliance may contribute to a reduction in access to acute care in California. SB 1953 raises costs for all suppliers in the general acute care market. Generally, rising costs pressure less efficient suppliers to leave markets, and there is evidence that this economic logic holds among hospitals. In fact, a California study found that hospitals facing a greater cost burden associated with SB 1953 were more likely to close.[131] In line with the analysis of Chapter 6 of this report, a study by Scheffler finds that financial performance significantly predicted hospital closures in California between 1995 and 2001.[132] Anecdotally, several hospital representatives we interviewed for this study indicated that they may close rather than pay the costs needed for SB 1953 upgrades. In addition, news stories note potential closings related to SB 1953, although we caution that policy compliance costs may be only one of several factors driving these decisions.

Distributional Effects and Equity

Hospitals bear all the initial outlays for SB 1953 projects under the status quo.[133] Other things being equal, financially stressed hospitals are more at risk for exiting the general acute care market in response to the policy. The analysis we provide on the affordability of SB 1953 in this report estimates that the share of hospitals in severe financial distress will increase by 18 percentage points (to 40 percent total), although some types of hospitals will face greater challenges than others. Our estimates indicate that CAHs and public health care district hospitals will be most stressed by SB 1953 compliance requirements: In our model, the number of hospitals in financial distress in these categories more than doubles after accounting for estimated compliance costs (see Figure 6.2). We estimate the cost for retrofitting this subset of hospitals at $3.4 billion (see Table 6.5).

Feasibility

As the status quo policy, the state has already approved the standard, and hospitals have already begun to respond to the policy by planning or completing retrofits or by preparing for hospital conversions or closures. Existing reporting mechanisms in place allow OSHPD to monitor progress toward SB 1953, and the rate of compliance is currently 17 percent of hospitals in the state. Although the policy does put the onus on hospitals to estimate retrofit costs and weigh business alternatives, these information gathering activities do not differ substantially from other capital planning activities for hospitals.

---

131 Tom Chang and Mireille Jacobson, "What Do Nonprofit Hospitals Maximize? Evidence from California's Seismic Retrofit Mandate," unpublished manuscript, 2012.

132 Richard Scheffler, *California's Closed Hospitals, 1995–2000*, Berkeley: Nicholas C. Petris Center on Health Care Markets and Consumer Welfare, University of California, Berkeley School of Public Health, 2001.

133 Consumers, however, may shoulder some of these compliance costs as prices adjust—although the ability to adjust prices is often limited by payers. See Chang and Jacobson, 2012, for a more detailed discussion.

Other Considerations

Because SB 1953 mandates significant expenditures for hospital construction, it may have an impact on the construction market in California as a whole. In brief, SB 1953 creates an increase in the demand for hospital construction, which tends to increase both the quantity and price of hospital construction in the market. Because the hospital construction market is part of the larger construction market, these higher prices affect all market participants, generating social costs in excess of the direct costs paid by hospitals for SB 1953. Measuring these costs is beyond the scope of this study.

Alternative Financing Models

One way to reduce the impact of seismic events on patients and encourage more hospitals to invest in seismic improvements is to lower the costs of SB 1953 to hospitals by spreading them more broadly among beneficiaries of the policy. Arguably, any person near the epicenter of a future earthquake in California stands to benefit directly from the availability of acute care, and people in the state as a whole benefit from the certainty that hospitals will continue to function after an earthquake, no matter where in the state they might be at the moment. Although policymakers would need to weigh the best use of public funds, SB 1953 does have clear social benefits and may merit public support.

A number of alternative funding models could be adopted to lower the costs hospitals face and thus encourage greater investment and reduce financial distress for some hospitals. The models fall into two broad categories: (1) direct grants to support seismic improvements and (2) programs to reduce the cost of financing seismic improvements.

Directing grants to support seismic improvements would require the state to allocate funds that could be provided to hospitals to help offset some of the costs of retrofits or new construction. This type of funding has been made available in other states to facilitate seismic improvements. For example, since 2010, Oregon has funded the Seismic Rehabilitation Grant Program, which is a competitive grant program that provides funding for seismic rehabilitation of critical public buildings, including schools, acute care hospitals, and fire stations.[134] California offers small grants to homeowners for seismic retrofits through the Earthquake Brace and Bolt Program.[135] At the national level, FEMA has a number of grant programs for states to use to improve earthquake preparedness.[136] Funding for a grant program might come from the general fund, by issuing bonds, or from the federal government through FEMA grants to states to improve preparedness for natural hazards. For example, Proposition 4, passed in November

---

[134] Business Oregon, "Seismic Rehabilitation Grant Program," webpage, undated.

[135] See California Residential Mitigation Program, Earthquake Brace and Bolt Program, homepage, undated.

[136] See FEMA, "Earthquake Grants," webpage, last updated June 14, 2018.

2018, authorized the state to issue $1.5 billion in bonds to support grants to hospitals for the construction, expansion, and renovation of children's hospitals in California.[137]

A grant program could be made available to all hospitals or could be targeted in ways to address the biggest barriers to seismic safety. For example, the program could target hospitals in regions that are particularly vulnerable, are the least prepared, or are in financial distress.[138] Also, the program could be used to support retrofits, new construction, or both.

Reducing the cost of financing investments in seismic retrofits and new construction is another way to reduce the costs to hospitals and facilitate greater seismic safety investment on their part. One option would be to expand access to public-sector bond funds to nonpublic hospitals. Other alternatives are loan insurance programs and loan guarantees that can reduce the cost of capital, particularly for hospitals that may have some financial difficulties and have lower credit ratings. For instance, OSHPD currently operates a loan insurance program for hospital construction called Cal-Mortgage.[139] Cal-Mortgage loans are guaranteed by the "full faith and credit" of the State of California, which allows borrowers with weaker credit ratings to receive a lower interest rate on the loan. The program is not specific to seismic improvements and is not available to private for-profit hospitals. However, the state could consider whether the program could be expanded.

Effectiveness

Changing the financing model does not necessarily change the requirement that all hospitals meet SB 1953. A grant program or loan guarantee could be added to the existing legislation. Hospitals could still be required to make seismic improvements by a particular deadline, but they could be eligible to receive some financial assistance. This assistance would make seismic projects more affordable, potentially increasing the rate of compliance relative to the status quo, reducing the risk of closures, and likely increasing the speed at which improvements occur.

Another consideration that might bear on effectiveness is that OSHPD's current loan insurance program, Cal-Mortgage, has, to date, not been used in any substantial way by hospitals to help finance their seismic improvements,[140] which raises the question about uptake of an

---

[137] For example, see Ballotpedia, "California Proposition 4, Children's Hospital Bonds Initiative (2018)," webpage, undated.

[138] However, policymakers would need to consider mechanisms to ensure that doing so does not create a moral hazard problem by disincentivizing their own investments.

[139] OSHPD, "Construction Financing: Cal-Mortgage Loan Insurance Program," webpage, undated. A similar program, the California Capital Access Program Seismic Safety Financing Program, provides loan insurance to small business owners and homeowners to reduce the costs of borrowing to make seismic improvements. California Pollution Control Financing Authority, "California Capital Access Program (CalCAP) Seismic Safety Financing Program," webpage, undated.

[140] As of 2016, OSHPD's Cal-Mortgage program insures 91 projects, 18 of which belong to hospitals. The original principal insured for hospitals totaled approximately $660 million, of which $576 million remained outstanding. Given that data gathered for this research show that a single hospital retrofit project has costs on the order of $100 million, the uptake rate of the program appears low relative to the total capital expenditures associated with SB 1953 compliance.

expanded program. As we noted, Cal-Mortgage is presently available to public and not-for-profit hospitals. The program has a total loan balance capacity of $3 billion, $1.7 billion of which was obligated as of November 2017.[141] The most creditworthy hospitals in this group do not need the loan insurance that Cal-Mortgage offers to obtain favorable capital rates. On the other hand, the program cannot accept all risks: It has an application process designed to assess an applicant's creditworthiness before extending loan insurance. In other words, although Cal-Mortgage improves access to capital to the extent possible by pooling risk, it cannot do so in disregard of the costs of defaults to taxpayers.

Cost

Relative to the status quo, the overall cost of meeting the SB 1953 requirements under alternative financing models depends on the exact type and rate of subsidy provided and the eligible set of projects targeted. If the alternative financing model encourages greater compliance, total direct costs incurred to society would very likely exceed the costs of the status quo because more hospitals are retrofitted or rebuilt. In particular, a program that solely targeted hospitals that planned to close would add to the total cost of compliance, because public funds would be added to the private funds spent on the status quo level of compliance. At the same time, because fewer hospitals would close, such a policy would maintain greater access to acute health care supplied both before and after an earthquake and reduce pressure on health care costs relative to the status quo. It is beyond the scope of this analysis to weigh the relative value of these costs and benefits or to determine the impact of varying levels of subsidy support on compliance.

Distributional Effects and Equity

Under alternative funding models, the distribution of costs would change so that the state bears some portion of the costs either through direct funding to hospitals or through loan guarantees. How the reduction in costs to hospitals is distributed depends in large part on the structure of the program. If the program targets specific geographic areas or hospitals with poor financial health, the benefits to hospitals would not be evenly distributed.[142] On the other hand, this kind of approach may succeed in avoiding a greater number of hospital closures per dollar spent. Before embarking on an analysis of the relative trade-offs between these costs and benefits, policymakers would need to provide some guidance about which hospitals or projects are worth public support and the acceptable level of risk for hospital closure.

---

[141] OSHPD, *Cal-Mortgage Loan Insurance Division Monthly Activity Report Ending November 30, 2018*, Sacramento, 2018.

[142] More generally, this type of program would benefit only hospitals that have not yet met the requirements of SB 1953. Some might argue that such a policy is unfair because it rewards hospitals that delayed compliance and does nothing to help those that complied in a timely way.

Feasibility

Implementing a grant program requires the development of infrastructure and processes for soliciting and selecting grantees, implementing awards, and monitoring progress and performance. There are a number of existing models that could be adapted and expanded to create the infrastructure and processes needed to implement such a program. The challenge is in finding and allocating the money at the state level to fund the program and determining the amount of funding required to induce compliance for the targeted hospitals. As noted above, such programs could be made available to all hospitals or could be targeted to achieve specific policy objectives or maximize the public good. For example, an independent commission could be appointed to allocate available funds in a way that prioritizes maintenance of regional capacity for critical services. In such a model, hospitals in financial distress might be prioritized for funding if the communities they serve would be otherwise vulnerable. Distressed or poorly prepared hospitals in markets with excess capacity might be given a lower priority.

Other Considerations

Relative to the status quo, alternative financing may increase the impacts of seismic compliance on the construction market if this financing increases the number of hospitals that ultimately comply rather than close.

Regionalization

Earlier in the report, we saw that the burdens of the SB 1953 requirements and the 2030 deadline are not felt equally among the state's hospitals. Seismic upgrades for some hospitals involve considerable additional expense but fit reasonably well into timelines for other capital upgrades. For other hospitals, compliance with SB 1953 requires wholly separate capital projects that would not likely be undertaken at all except for the law. But regardless of cost, SB 1953 applies to all hospitals.

One alternative, discussed in part by Chang and Jacobson,[143] would involve replacing the requirement that all hospitals be compliant with one that seeks to ensure that there is a *sufficient* number of compliant beds in a *region*. Operationally, after a major earthquake, additional capacity at compliant hospitals would compensate for lost capacity at noncompliant, possibly damaged hospitals, until the latter could be repaired. For example, as noted in Chapter 1, even though 8 of 91 acute care hospitals in Los Angeles County were evacuated after the Northridge earthquake, regional capacity proved adequate to absorb this loss. On an ongoing basis, hospitals with lower compliance costs would provide a greater share of a region's supply of compliant beds, thus reducing the overall costs to the region and the state. Over time, policymakers could increase or lower the required number of beds per region in response to changing demands and standards.

---

[143] Chang and Jacobson, 2010.

This policy could be implemented in one of three basic ways. First, regional allocation could be centrally determined by an independent commission at the state level based on input from regional stakeholders. This would afford the state the most discretion in setting compliance goals. Although this approach would not preclude hospitals from choosing to exit the general acute care market, policymakers could nonetheless take that information into account as they weighed their plans. At the same time, however, such a process may necessarily give hospitals an incentive to overestimate compliance costs.

Second, hospitals could bargain with each other through some sort of regional governance body, such as the existing health care coalitions promoted under the Hospital Preparedness Program cooperative agreements (within the Office of the Assistant Secretary for Preparedness and Response, U.S. Department of Health and Human Services).[144] Coalitions typically involve some sort of organizational venue in which hospitals (including those from competitor provider networks) meet with nonhospital clinical providers, emergency managers, and other key stakeholders and could conceivably strike bargains in which noncompliant members compensate compliant members for providing the required number of compliant beds for the region. Such an approach could be guided by centrally determined regional targets established by the sort of expert commission mentioned above or could be determined regionally under some set of guidelines.

Under this approach, there could be bilateral trades, with some hospitals compensating others to provide seismically secure beds that could be used in the event of an earthquake. This bargaining would build on the practice of hospitals contracting with other nearby hospitals with low patient censuses to gain access to beds to accommodate patients during high-demand periods. Alternatively, coalitions could create a common fund for upgrades, with members who had made substantial prior investments in seismic compliance contributing less. An additional feature of the coalition approach is that it could involve consideration of the capabilities of nonhospital clinical facilities, which might be especially relevant after the initial acute phase of a seismic incident and might promote a broader "systems" approach to postearthquake health care that includes consideration of nonhospital resources.

As a third alternative, policymakers could create a market akin to a pollution cap-and-trade system to drive the allocation of regionally tailored compliance requirements. Instead of trading rights to pollute, participants in this system would trade rights to maintain *noncompliant beds*. After determining the sufficient number of beds for a given region, a central authority (such as OSHPD) could initially allocate permits for a region by auctioning them or distributing them among the participating members.[145] Hospitals would then weigh whether it is less expensive to

---

[144] See U.S. Department of Health and Human Services, "Hospital Preparedness Program (HPP)," webpage, last reviewed March 4, 2019.

[145] Although both of these options arrive at cost-efficient outcomes after trading, they differ in their distributional implications. Auctions capture rents for the state, while distribution allows the value of those rents to be captured by initial permit holders. Policymakers can, therefore, use initial allocations to address equity concerns and secure buy-in from market participants. In either trading system, regulators would need to consider the time horizon of the

purchase permits or pay to come into compliance. This system overall leads to an efficient allocation where compliant beds are provided at lowest cost.

Effectiveness

Regionalization would allow policymakers to set rates of seismic compliance that are less than 100 percent, thus reducing the number of compliant beds and the direct costs of implementing the policy. The impact of the reduction in the number of compliant beds on public safety could be reduced somewhat if compliant beds were concentrated in areas with high risk and high patient demand. It is conceivable that regionalization could speed compliance by easing costs, but such gains depend considerably on the overall targets set and other factors. Costs to hospitals would nonetheless remain significant, so the risk of closures would still persist in some cases. But on a region-by-region basis, state policymakers would have the opportunity to evaluate the potential impact of such closures and adjust local plans to best meet the state's and the community's needs.

Cost

By design, regionalization lowers compliance costs relative to the status quo because a smaller number of hospital beds would be required to be compliant. Under SB 1953 as currently implemented, all general acute care hospitals in the state must meet state seismic standards or exit the market. The flexibility that regionalization introduces specifically eases this constraint, imposing less than total compliance costs on hospitals. Whether through bilateral trade or permit markets, participants may further benefit from gains in cost efficiencies by retrofitting or building new capacity wherever it costs least to do so.

Distributional Effects and Equity

Regionalization affords policymakers two avenues for addressing equity issues. First, when policymakers set regional requirements, they can incorporate input from stakeholders to help determine likely impacts and make adjustments as desired. Second, under the permit market option, policymakers can initially allocate permits for noncompliant beds (which have value because of their scarcity) to hospitals perceived as disadvantaged or otherwise deserving of support.[146]

---

underlying rights and requirements. Would the rights be of a one-shot nature, with permanent structures devoted to meeting compliance in place by 2030, or would market participants need to acquire permits on an ongoing, periodic basis? If policymakers imagine that standards will continue to evolve, perhaps toward 100 percent compliance, they might reduce the cap for noncompliant beds over time and reissue or reauction rights periodically. The ability to transfer rights to noncompliant beds over time (*banking*) in this kind of system has further implications for the value of permits, cost of compliance, and the speed of compliance that lie beyond the scope of this report.

[146] In making such allocations, however, policymakers would need to be aware of some possible risks. Hospitals with the ability to bear compliance costs are likely already in strong financial condition, compared with hospitals with high compliance costs. Thus, absent any extra allocation to hospitals with high compliance costs, the policy may effectively transfer resources to hospitals that are already relatively more profitable. At the other extreme, if

Although most tradable-permit systems involve some initial, free allocation of permits to address equity goals, this facet of the policy introduces another complication to consider. To the extent that administrators rely on self-estimated individual compliance costs to determine how to allocate noncompliant beds, market participants would have incentives to bias their estimates. Trading would still lead to an efficient allocation over time, but the rents extracted by initial grantees could be substantial. Administrators could circumvent this concern by auctioning rather than giving away permits at the outset, although this option again would leave hospitals facing high compliance costs in the same straits they are in today.

In any case, tradable-rights externalities fundamentally allow participants to pay not to comply with regulations. Typically, participants with relatively more resources can afford to avoid more compliance. This disparity often forms the basis of ethical arguments against such policies and can impede public buy-in. Again, policymakers can use initial allocations to offset these concerns to an extent, but the inherent perception of the policy as "paying to break the law" can surface as an issue.

Feasibility

Regionalization would require addressing a host of complex implementation issues. Although hospital SPC and NPC designations already take into account underlying seismic risk, attention to the size and overall characteristics of the region served are not considered. Regionalization policies would first need to determine both the optimal size and scope of each region and the number of compliant beds that each region requires. Although the regional structure implied by the Hospital Preparedness Program's cooperative agreements might not be ideal for this purpose, it is nonetheless an existing structure that may serve as a solid starting point. Adopting this structure would lower administrative costs and leverage existing relationships to facilitate planning or trading for compliance.

However, it would be wise to engage in careful analysis and reflection before accepting any such structure. Providing a detailed guide to this analysis is beyond the scope of this report, but it would likely include mapping population exposure to expected damage patterns under various earthquake scenarios, with the objective of ensuring that a high percentage of residents in a region could access a compliant general acute care hospital within a defined time period. In addition, such analyses would ideally account for a variety of other critical physical and economic realities (e.g., road networks).[147] Policymakers could then define a "coverage target" for all regions that ensured equal access to compliant beds for all residents in the state and set

---

hospitals with low compliance costs were allocated all the permits they need (or more), they would, in effect, be able to "free ride" on the investments of other hospitals. Policymakers would need to strike the right balance to meet equity goals while ensuring adequate buy-in from market participants.

[147] For example, regional plans could incorporate analyses of road networks and potential postincident choke points (e.g., bridges vulnerable to collapse) to ensure access to care under critical scenarios. Economically, regions need to have enough participants to ensure competitive transactions. Sparsely populated regions with relatively few hospitals may have few opportunities for trade and may be subject to the exercise of market power.

regional boundaries that ensure such access, factoring in variations in demand and other considerations, as appropriate.

Once the overall level has been set, regionalization could proceed by either centrally determining allocations within the region (by a health care coalition or other similar entity) or allowing participants to determine the allocation of unsecured beds. In either case, regionalization would require a shift in OSHPD's regulatory role. Under this policy alternative, OSHPD (or another entity) would take responsibility for determining whether regional compliance goals have been met, requiring a finer level of site inspections.[148] With tradable rights, OSHPD (or another entity) would also need to record or centrally facilitate transactions that support compliance. Administrators would then need to spend resources to ensure smooth adoption and best use of the options this policy presents.

Under a bilateral trading system in particular, policymakers would need to give some attention to antitrust concerns. Hospitals within a region arguably hold considerable market power. Private meetings between these entities to bargain over compliance trades could easily touch on anticompetitive topics, such as capital planning and market strategies, raising concerns of collusion. State administrators could allay such concerns by providing the auspices under which participants meet and monitoring, if not moderating, their discussions.

Finally, we note that all regional approaches assume some fungibility of hospitals. Because of the limitations of some insurance and provider networks, however, patients might not necessarily be covered at some hospitals in a region. Cross-network patient transfers may further make it more difficult to maintain the integrity of patients' electronic medical records. Disaster plans and agreements often address these issues for short-term emergencies, but doing so over the longer term after a disaster may create additional challenges.

Other Considerations

Relative to the status quo, regionalization may reduce the impacts of seismic compliance on the construction market if it reduces the overall level of compliance required.

Timing Flexibility

Timing flexibility denotes the family of policy options that relaxes any of the deadlines for SB 1953 compliance. These policies could take a variety of forms. Extensions could be granted for hospitals based on their financial health or other characteristics. For example, as noted in the analysis in Chapter 4 on affordability, CAHs and high-Medi-Cal hospitals are disproportionally affected by SB 1953–imposed financial distress. Extending timelines for these groups could spread their financial burden over time. Any number of other possibilities in this vein exist. To illustrate, policymakers might

[148] Verifying whether compliance plans have been met and are indeed sufficient in the case of a seismic event may present a significant challenge. Policymakers and regulators would need to scrutinize the details of any such plans carefully and perhaps issue guidance on allowable elements.

- allow (some) hospitals to rebuild or retrofit in line with their normal capital replacement schedule
- customize deadlines based on project characteristics by type or complexity of the building project
- allow rolling deadlines for sequencing construction or retrofitting of multiple buildings on the same campus.

Effectiveness

Timing flexibility in general would delay full compliance for California hospitals. This policy would therefore reduce the level of compliance across the system in the short run as hospitals take advantage of extended deadlines. Nonetheless, timing flexibility does not change the final specified level of compliance for general acute care hospitals. This policy option is therefore at least as effective as the status quo after its final deadline passes. The cost to comply may still induce some hospitals to exit the general acute care market, but savings from timing flexibility would likely reduce this risk relative to the status quo.

Costs

Providing additional time for hospitals to achieve compliance with SB 1953 would offer some relief from the mandate's costs. In effect, hospitals could delay investments and leave their capital less constrained in the near term. Similar to alternative financing options, the total direct costs of meeting the SB 1953 requirements under timing flexibility depends on the exact nature of extensions available under the policy and the eligible set of projects targeted. On the other hand, because timing flexibility necessarily reduces compliance in the near term, these policies impose additional costs on society in the form of greater exposure to the risk of acute health care failures in the event of an earthquake. Determining whether the value of the extension to hospitals exceeds its cost to society is beyond the scope of this analysis.

Distributional Effects and Equity

Under timing-flexibility policies, society as a whole would accept some additional risk relative to the status quo, as fewer hospitals would be ready to meet SB 1953 requirements by the 2030 deadline. Hospitals would continue to bear all of the direct costs of compliance, with some passed through to consumers over time. The distribution of cost savings would depend directly on the structure of the program—if the program targets specific geographic areas or hospitals with poor financial health, for example. More generally, this type of program would benefit only hospitals that have not yet met the requirements of SB 1953. This disparity can again affect buy-in; these policies may be viewed as unfair because they reward hospitals that delayed compliance and do nothing to help those that complied in a timely way.

Feasibility

Because the deadlines of SB 1953 are written into the law, California would need to adopt new legislation to provide timing flexibility. Policymakers would need to determine how much time hospitals targeted under this option would require to comply. Alternatively, policymakers could instead determine the maximal time they are willing to extend hospitals to retrofit for SB 1953.

Other Considerations

Timing-flexibility options have at least two implications beyond hospitals. First, extended deadlines to allow hospitals more time to achieve compliance spreads demand for construction across time and reduces pressure on prices, both for hospitals and the market at large. Second, as desirable as providing flexibility may be for hospitals, adopting this alternative may negatively affect policy credibility. Continuing revisions to regulatory deadlines may lead hospitals to expect that future deadlines are also malleable, potentially inducing a noncompliance trap. Policymakers considering this option may weigh adopting mechanisms to ensure that hospitals view any extension as final and binding.

Finally, policymakers considering timing-flexibility options should bear in mind that changes in economic conditions may inadvertently diminish their impacts. Hospitals mainly benefit from these policies by delaying costly investments. If prices for compliance rise faster than expected, however, these savings might not materialize. Although such changes may arise as a product of market forces, they can also be influenced by local ordinances or state policies.[149]

Streamlining OSHPD Approvals

Stakeholders interviewed for this study frequently identified regulatory approvals through OSHPD as a significant contributor to project timelines and costs, drawing comparisons with other states with faster approval processes. Measuring the impact that the approval process has on hospital construction costs overall is beyond the scope of this study, but we found no further external data or analysis to corroborate this position. Nonetheless, because of the saliency of this idea, we examine the potential relief in compliance costs for hospitals that may arise from streamlining OSHPD approvals through some to-be-determined, specific set of policies or through an increase in OSHPD staff and capacity.

---

[149] For example, in 2015, the City of Los Angeles adopted the Non-Ductile Reinforced Concrete Ordinance that requires some 1,500 commercial buildings erected before 1977 to demonstrate compliance, complete a set of retrofits (by 2040), or set a schedule for demolition. This ordinance will likely increase demand in the local construction market in the 2030–2040 interval. See Seismic Ordinances of California, "Los Angeles: Non-Ductile Concrete Structures," webpage, undated.

Effectiveness

In theory, improving the approval process or increasing the number of administrators or inspectors at OSHPD could reduce construction project timelines and thereby reduce costs to hospitals by reducing project escalation costs and risks related to schedule slip. In practice, the effectiveness of such a policy depends on the expected time saved per project. Estimating potential time saved presents substantial challenges and makes calibrating this policy to achieve a desired level of response highly impractical.

The difference in California's administrative costs may come from administrative processes, the state's capacity, and the complexities associated with implementing national building code standards within the context of the geohazard conditions. Streamlining review processes would address only one of those three. During interviews with OSHPD staff, respondents pointed to several efforts over the years to streamline the process, including

- **collaborative review:** This "fast-track" process for large, complex projects allows regulators and hospitals to negotiate the timeline up front. According to OSHPD, the agency meets its commitments under this policy 90 percent of the time.
- **e-plan review:** This process allows projects of any size and dollar value to be submitted electronically. In the past, e-plan review did not cover all projects and had limited capabilities. Under this expansion, OSHPD is increasing the availability of online drafting boards. These allow regulators to interact with hospitals on plans electronically and reduce the need for travel and project timelines.

OSHPD staff further asserted that a significant share of timeline delays arises because of hospitals or the construction firms they work with. If true, the impact of process streamlining might be limited. The sort of detailed analysis needed to shed analytic light on these issues was beyond the scope of this study but should be investigated further before any decisions about process reform are taken.

Last, although we cannot independently validate the statement, OSHPD reported no significant staff shortages at present. Given the coming wave of compliance construction and need for administrative approvals, OSHPD may have further staffing needs in the future. In a discussion of this question, OSHPD staff noted that they are aware of the trend and can hire external contractors to deal with high demand in the short run, should their staffing plans prove insufficient to needs.

Cost

Potential benefits notwithstanding, streamlining processes where possible could result in net efficiency gains, in terms of both cost savings for hospitals and administrative cost savings for OSHPD itself. Likewise, adding administrators, inspectors, or other employees may also result in net gains if the savings in project timelines exceed the compensation paid. In terms of direct project costs, reduced timelines typically result in lower costs for all of the following: temporary services, risk and contingency, escalation, and security. Because subcontractor and general contractor profit are typically a percentage of the total contract, these costs would fall as well.

Additionally, hospitals would benefit by spending less time shut down for retrofits or by bringing new hospitals online sooner.

Feasibility

The fact that OSHPD has recently undertaken other process changes demonstrates that some critical processes can be improved without new legislative authority. Absent a specific set of policy proposals, however, we cannot determine administrative feasibility of this option.

Distributional Effects and Equity

In general, improving administrative efficiency would benefit hospitals that have not yet submitted SB 1953–related building proposals to OSHPD. It would make little sense to target such policies for specific types of hospitals or projects unless they ordinarily require high effort, such as the complex projects targeted by collaborative review. Similar attention to small projects—as would typically come from rural hospitals, for example—would result in little net gain of time and therefore little net benefit to society.

Other Considerations

To the extent that streamlining OSHPD processes goes beyond SB 1953 and reduces administrative costs for hospital construction generally, policies in this vein are likely to reduce costs more broadly and thereby encourage new construction in the market.

Discussion

In this chapter, we considered four broad categories of alternative policy options that could be used to achieve the goals of SB 1953 to "have general acute care hospital buildings that not only are capable of remaining intact after a seismic event, but also capable of continued operation and provision of acute care medical services after a seismic event."[150] Within each broad category, we describe a range of possible policy options that could be considered (see Table 8.1 for an overview). Still, the options we considered are intended to be illustrative of different types of available policy levers and are not intended to be exhaustive.

Looking across the four alternatives, none clearly dominates the others. Rather, each alternative has its advantages and disadvantages relative to each other and to the status quo. The relative magnitude of the advantages and disadvantages, however, is difficult to assess at this point because it would depend heavily on how the policy was designed. We have highlighted many of the key design considerations and how they might affect the effectiveness, cost, feasibility, and distributional effects and equity in the discussion of each alternative but cannot be more precise without specific design decisions.

---

[150] OSHPD, 2015.

More generally, as noted, there are still a number of big-picture questions that should be considered before making any policy decision. These include whether the policy, given the nature of health care delivery, should look beyond hospitals to include the broader health care system; how the costs associated with implementing SB 1953 relate to the benefits of seismic resilience; and whether there are alternative public health investments that would provide a greater benefit to society.

Table 8.1. Policy Options Considered

Policy Group	Specific Options
1. Status quo	—
2. Alternative financing	Direct grants to support seismic improvements
	Reduce the cost of financing investments in seismic retrofits and new construction
3. Regionalization	Centralize determination
	Use bilateral trading system
	Permit trading for noncompliant beds
4. Timing flexibility	Customize deadlines based on hospital characteristics
	Allow (some) hospitals to rebuild or retrofit in line with their normal capital replacement schedule
	Alternatively, customize deadlines based on project characteristics by type or complexity of the building project
	Alternatively, allow rolling deadlines for sequencing construction/retrofitting of multiple buildings on the same campus
5. Streamlining OSHPD approvals	Improve approval process
	Add staffing

9. Summary of Findings and Recommendations

After 25 years of implementation, hospital owners, regulators, workers, and patients agree that the intent of the objectives of SB 1953 continue to have merit. The original legislation was passed in an attempt to enhance the resilience of California's hospital system to seismic events to (1) ensure the safety of patients and health care providers and (2) enable the uninterrupted delivery of health care services to Californians. There is little question that it is in the public interest to have seismically resilient hospitals. However, given that California's SPC-1 buildings—those at risk of collapse—will have been retrofitted, replaced, or taken out of service by 2020 or shortly thereafter, there is now an opportunity for legitimate debate regarding how to most effectively and efficiently enhance resilience in future years. Although inpatient, acute care hospitals continue to play a critical role in health care delivery, the long-term trend in the industry has been toward greater reliance on outpatient services. It is important that capital investments in hospital facilities are aligned with current models for service delivery. It is therefore appropriate at this point to reassess whether SB 1953 remains the best policy to ensure resilience and to consider potential alternatives.

To this end, this report provides an up-to-date assessment of the cost implications of SB 1953 implementation for California hospitals and, indirectly, the communities they serve. This assessment includes some elements similar to prior RAND studies on the costs of compliance,[151] yet it is fundamentally different in that it estimates costs specifically for California's portfolio of SPC-2 buildings and noncompliant NPC buildings. By placing cost estimates of SB 1953 in a larger policy context, this study provides a foundation for a discussion of how policy can effectively ensure the seismic resilience not just for California's hospitals but the state's health care delivery infrastructure more broadly.

Findings

As seismic compliance efforts are reoriented toward the 2030 deadline, hospitals and regulators are likely to learn rapidly about the costs of future compliance. Since 1994, a broad range of legislative and regulatory decisions have shaped SB 1953 implementation. These decisions have generally benefited hospitals by extending the time available to achieve compliance or providing greater flexibility in how to achieve compliance. Given the number of remaining noncompliant SPC-2 buildings in the state, a large number of seismic projects must be planned and implemented in the near future. Although some of this planning is already underway, new information will become available on compliance costs as more projects are designed, enabling more-constrained estimates of total costs. Meanwhile, hospitals and

[151] Meade, Kulick, and Hillestad, 2002; Meade and Kulick, 2007.

regulators implementing SB 1953 will continue to be influenced by future innovations in policy design, engineering, and health care delivery. Therefore, further changes in SB 1953 or its implementation may be warranted to maintain the reasonableness of the requirements, take advantage of new opportunities, and reduce unintended consequences of the policy.

Hospitals evaluate capital investments in seismic compliance in the larger context of individualized business planning, making it difficult to anticipate their actions and costs. Hospitals make decisions based in response to community needs, service delivery, financial health, market conditions, and competition. These factors vary widely from one hospital to another. Past studies of the costs of SB 1953 by RAND assumed that hospitals would uniformly opt to replace noncompliant buildings because of the disruption caused by invasive retrofit actions.[152] In this study, however, we discerned no general rules or assumptions that can be used to anticipate whether hospitals will retrofit or replace noncompliant buildings or take those buildings out of service. Although replacement has been pursued by some hospitals, retrofitting existing buildings is more cost-effective for others. Many hospitals are still in the process of evaluating the best course of action with respect to compliance with the 2030 deadline.

SB 1953 requires hospitals to make capital investments they would not otherwise make or at a time they would not otherwise choose. SB 1953 was intended to accelerate the recapitalization of aging, noncompliant hospitals through seismic retrofits, new construction, or removal of old buildings from service. Requiring hospitals to link the timing of these actions to imposed deadlines in 2020 and 2030 reduces the flexibility of hospitals to make capital investments when and where they will be most advantageous. Synchronizing capital investments in seismic resilience with other capital investments could help achieve resilience at lower cost, although this could slow the rate at which compliance is achieved.

Because the original deadline for addressing SPC-1 buildings was 2008, the challenge of achieving compliance has clearly resulted in changes to deadlines and the granting of extensions. Although hospitals have used the extra time to complete necessary retrofits and new construction, the overall cost of compliance with the 2030 deadline remains high for many hospitals. As the 2030 deadline approaches, regulators will likely need to continue to weigh enforcement of the regulation against its impacts on hospital access and capacity.

The lack of prior cost-benefit analyses of SB 1953 means that decisionmakers do not fully understand the trade-offs the policy creates for the health and well-being of Californians or the hospitals that serve them. The consequences of SB 1953 affect the entire state and could potentially result in reductions in hospital capacity. The trend toward greater reliance on outpatient services and facilities, when combined with the implications of retrofitting existing buildings or replacing those buildings, could result in a net reduction in hospital capacity in California by 2030. However, evidence of reductions in capacity are based on a small number of hospitals; therefore, we cannot estimate the scale of such reductions or whether they would be offset by expansions of capacity in other hospitals. Meanwhile, developing a comprehensive understanding of the effects of SB 1953 necessitates consideration of the benefits of the policy—

---

[152] Meade, Kulick, and Hillestad, 2002; Meade and Kulick, 2007.

an aspect of SB 1953 for which there is a significant knowledge gap. Filling that gap was beyond the scope of this study, but such benefits should nevertheless be considered in deliberations regarding future policy changes.

The costs of future compliance with SB 1953 are large in aggregate but will vary significantly among hospital systems and individual buildings. As hospitals proceed with planning and design efforts around future capital investments in seismic safety, a clear understanding of the costs of SB 1953 will emerge. Although a number of hospitals have undertaken such work, many have not. Based on the analysis of seismic project cost estimates provided to RAND and extrapolated to California, the total costs of compliance are estimated at \$34 billion–\$143 billion by 2030. The low value assumes that all SPC-2 hospitals retrofit to SPC-4D, while the higher value assumes that all hospitals replace their SPC-2 buildings with new construction. However, because hospitals will pursue a mix of retrofit and new construction, neither value is realistic. Rather, the actual cost will fall somewhere within that range. These estimates include costs associated with NPC compliance, as well as future cost escalation, but do not include investments that have already been made since 1994 to address noncompliant SPC-1 buildings. More routine and detailed collection of information on compliance strategies, expenditures, and characteristics of capital investments in noncompliant buildings would improve the ability of hospitals and state regulators to monitor compliance and improve cost estimation.

The costs of implementing seismic compliance construction projects have increased in recent years at a rate faster than inflation. While prior RAND studies estimated the real rate of increase in construction costs as similar to the background inflation rate or consumer price index, the rate of increase in construction costs has outpaced inflation for more than a decade. The California consumer price index has averaged a 2.3 percent annual rate of increase since 2002. Yet recent, short-term estimates of the annual increase in general U.S. commercial construction costs and California construction costs range from 4 percent to over 7 percent per year, respectively. This increase is a function of multiple factors, including rising construction and labor costs in the state, increasing costs for nonseismic code requirements, and fluctuating capital costs. Many hospital representatives interviewed for this study believe that cost escalation may accelerate in the future as hospitals compete for qualified contractors and labor to comply with the 2030 deadline. Whether such acceleration will in fact occur, however, remains uncertain. Nevertheless, these rates of increase drive up the costs of complex, multiyear hospital construction projects.

A number of factors influence the affordability of SB 1953, and affordability varies among hospitals. The affordability of seismic compliance is a function of both the capacity of hospitals to pay for seismic upgrades and their willingness to make those investments. Hence, different hospitals will have different perceptions of what is or is not affordable, as will other stakeholders. Some hospitals appear to have a sufficient combination of cash reserves, revenue, or access to financing to undertake seismic compliance projects. Nevertheless, financial health does not necessarily mean that seismic upgrades are affordable, because hospitals may have other capital investment needs and priorities. Meanwhile, approximately 34 percent of California's hospitals were identified as having at least some financial stress based on current financial data. Twenty-two percent were identified as experiencing severe financial stress. After including in the analysis of financial stress the potential costs of SB 1953 compliance that individual hospitals might incur, the proportion of hospitals with some or severe financial stress increased to 52 percent and 40 percent, respectively. Because of the social benefits they provide, hospitals often persist for extended periods with some degree of financial stress. Nevertheless, significant increases in financial stress are inconsistent with the long-term financial sustainability of hospitals.

Moreover, SB 1953 disproportionately affects certain categories of hospitals. The largest increases in financial stress were observed for CAHs, public health care district hospitals, and stand-alone hospitals that are not part of a larger hospital network. This reflects the disproportionate vulnerability of these facilities to the costs of compliance. These hospitals often serve a relatively large proportion of Medicare and Medi-Cal patients and have limited options for private financing of capital investment. Given that Medicare and Medi-Cal reimbursements have failed for years to keep up with inflation despite rising costs for the provision of health care services, SB 1953 creates significant long-term financial obligations while operating margins are shrinking.

The lack of public funding to incentivize implementation of SB 1953 overlooks the public-good benefits of hospitals, generally, and seismically resilient hospitals, specifically. Policymakers and the public at large recognize health care as one of the nation's critical infrastructure sectors: It is as essential to human well-being and safety as energy, water, or transportation infrastructure. Efforts to enhance seismic resilience of these other critical sectors have often been incentivized through public investment and subsidies. In part, the public ownership of such assets (e.g., roads, bridges, and dams) provides a mechanism for funding such investments. The lack of public subsidies to incentivize SB 1953 compliance stands in contrast to the role public funding has played in ensuring the resilience of other critical infrastructure.

A range of alternative policy frameworks could be deployed to help achieve seismic resilience for California hospitals. Potential policy proposals include public subsidies or loan guarantees to improve access to capital, right-scaling seismic requirements within regions to balance costs and benefits, continuing to provide flexibility in the timing of compliance, and building capacity in OSHPD to streamline compliance. Each approach is associated with different trade-offs in terms of safety, feasibility, complexity of implementation, and cost distribution. Furthermore, these approaches represent just some of the potential policy alternatives that merit consideration and debate.

Recommendations

This report does not make specific policy recommendations regarding whether SB 1953 should continue to be implemented as is, altered to change how the costs of seismic resilience are distributed throughout society (e.g., use of public subsidies), or modified to adopt a higher tolerance for risk. Nevertheless, in conducting this study, a number of opportunities for building resilience in California hospitals, independent of seismic upgrades, emerged.

Addressing long-standing knowledge gaps associated with the benefits and costs of SB 1953 implementation can provide important information to decisionmakers regarding the merits of pursuing future changes in seismic requirements. Analyses of the implications of SB 1953 have largely focused on quantifying the direct costs of compliance. Meanwhile, other considerations relevant to evaluating the value of the legislation have received comparatively little attention. For example, although there is general agreement that having seismic-resilient hospitals conveys benefits to Californians, no formal analysis of the benefits of SB 1953 has been conducted. Are the benefits of seismic compliance limited only to the avoidance of damage or disruption in hospitals in the event of a seismic event, or are there ancillary benefits for hospitals that have been overlooked? Similarly, how do the seismic requirements convey benefits to communities? As we have acknowledged, SB 1953 is likely also associated with significant opportunity costs. Such opportunity costs have not been analyzed or estimated. Improving our understanding of these costs and benefits would provide a more comprehensive understanding of SB 1953 and its net implications for California.

The more-detailed reporting by hospitals to OSHPD regarding seismic projects, combined with independent validation of project costs, would enhance the capacity of OSHPD and other actors to estimate costs and monitor capital investments in seismic construction projects over time. Previous RAND studies on the costs of compliance with SB 1953 were informed by hospital surveys conducted by OSHPD, which resulted in standardized information on the characteristics of noncompliant buildings (e.g., number of square feet, number of beds) that were integrated with cost metrics.[153] Even on a voluntary basis, routinizing such surveys and enhancing the collection and validation of cost information for seismic projects would help build a more comprehensive database of costs that could prove beneficial to hospitals planning seismic projects, as well as regulators and policymakers.

Enhancing pathways of communication between hospitals engaged in the planning and implementation of seismic projects and OSHPD can build understanding of compliance challenges and facilitate the identification of mechanisms for enhancing regulatory flexibility. In recent years, OSHPD has worked closely with hospitals to maintain compliance with legislative requirements of SB 1953 while pursuing opportunities to address common barriers to compliance. Such collaborative relationships between OSHPD and hospitals enable ongoing deliberation regarding the reasonableness of seismic requirements. This deliberation enhances the likelihood of identifying policy innovations that reduce the burden, and therefore increase the likelihood, of compliance.

---

[153] Meade, Kulick, and Hillestad, 2002; Meade and Kulick, 2007.

Appendix A. Definitions of Hospitals

The various types of hospitals used in this study are summarized in Table A.1.

Table A.1. Descriptions of Types of Hospitals Used in This Study

Type of Hospital	Description
Academic hospital	A hospital that is affiliated with a public or private university
Acute care hospital	A hospital that provides 24-hour inpatient care, including multiple basic health care services, such as medical, nursing, surgical, anesthesia, laboratory, radiology, pharmacy, and dietary services
CAH	A hospital that is located more than 35 miles from another hospital, has 25 or fewer acute care beds, and provides 24-hour emergency services, and where patients maintain an annual average length of stay of 96 hours or less
District hospital	A public hospital organized by cities and townships within a specific geographic area to provide hospital services, often financed through local tax revenue
High-Medi-Cal hospital	A hospital with a disproportionate number of Medi-Cal patient-days, suggesting a population with a greater proportion of lower-income patients
MHS	A hospital system composed of two or more general acute care hospitals
Private hospital	A hospital that is funded and operated by a private organization (either for profit or not for profit)
Public hospital	A hospital funded and operated by a government, such as a city, county, or district

Appendix B. The Alfred E. Alquist Hospital Seismic Safety Act and Subsequent Policy Changes

Hospital seismic safety legislation in California has its origins in the 1973 Alfred E. Alquist Hospital Facilities Seismic Safety Act (SB 519). However, a broad range of legislation has been implemented in subsequent years. This includes SB 1953, passed in 1994, as well as a number of subsequent changes. These various legislative actions are summarized in Table B.1.

Table B.1. Summary of Past California Seismic Safety Legislation

Legislation	Description	Year Enacted
SB 519	The Alfred E. Alquist Hospital Facilities Seismic Safety Act of 1973 (SB 519) established a seismic safety building standards program applicable to hospitals built on or after March 7, 1973, to be overseen by the Office of Statewide Health Planning and Development (OSHPD). SB 519 established the Hospital Building Safety Board to advise OSHPD on the implementation of the Alquist Seismic Safety Act and act as a board of appeals for hospital facilities in regard to seismic safety and fire and life safety issues.	1973
SB 961	This legislation transferred the authority of construction plan review of health facilities from local building departments to OSHPD. The Facilities Development Division of OSHPD was created to handle these new responsibilities. By 1991, construction observation activities (i.e., inspection) were added to OSHPD's duties, making it the single authority and enforcement agency on construction projects of health care facilities.	1983
SB 1953	This legislation mandated that all general acute care hospital buildings be assigned a structural performance category (SPC), ranging from SPC-1 to SPC-5. The law originally required that all SPC-1 buildings be removed from providing general acute care services by January 1, 2008, and all SPC-2 buildings be removed from providing general acute care services by January 1, 2030.	1994
SB 1801	This bill authorized OSHPD to extend the January 1, 2008, deadline for certain hospital buildings of a general acute care hospital if the hospital agreed that, on or before January 1, 2013, designated services would be provided by moving into an existing conforming building, relocating to a newly built building, or continuing in the retrofitted building that is in compliance with designated structural and nonstructural performance categories.	2000
SB 2006	This bill authorized any hospital, with regard to a general acute care hospital building located in seismic zone 3, to request an exemption from certain nonstructural requirements if the hospital building complies with certain year 2002 nonstructural requirements.	2000
AB 2194	This bill provided for the waiver of a general acute care hospital from certain standards with respect to relocating services on an interim basis as part of its approval plan for compliance with the act.	2000
SB 1661	This bill authorized OSHPD to grant an additional extension of up to two years after the January 1, 2013, deadline if the hospital building subject to the extension met prescribed requirements, including that it was under construction at the time of the request for the extension and that the hospital had made reasonable progress in meeting the deadline, but for reasons beyond its control could not meet the deadline.	2006

Legislation	Description	Year Enacted
SB 306	This bill authorized certain hospital owners who did not have the financial capacity to bring certain buildings into compliance by 2013 to instead replace those buildings by January 1, 2020, by filing a declaration to that effect that includes specified financial information and a fee to cover the additional costs.	2007
SB 499	This bill permitted a hospital to receive an additional two-year extension to the January 1, 2008 deadline on top of the existing two-year extension. It also reopened the window for hospitals to qualify for the up-to-two-year extension available to hospital buildings that filed building plans, submitted a construction timeline, and were under construction. This new option was available for hospitals that sought but did not receive reclassifications of their seismic status under Hazus. SB 499 also created an exemption from NPC-3 requirements, allowing hospitals to address those requirements at a future time period in conjunction with other improvements (such as SPC upgrades).	2009
SB 608	This bill authorized OSHPD to grant a three-year extension of the existing five-year extension. It granted an additional extension of up to two years if specified criteria were met. The bill also required a hospital owner that applies for an extension pursuant to this bill to pay to the office a fee for the costs of reporting required for this extension.	2010
SB 90	This allowed hospitals that have received extensions to January 1, 2013, of the January 1, 2008, seismic deadline for their SPC-1 buildings to request an additional extension of up to seven years. OSHPD was allowed to grant the extension if the hospital met several interim deadlines and requirements or if the hospital building was reclassified to a higher seismic status, allowing it to operate beyond 2013.	2011
AB 2557	This clarified that a general acute care hospital building may be used for purposes other than nonacute care hospital purposes if an extension of the January 1, 2008, deadline has been granted and, before the end of the extension, a replacement building has been constructed or a retrofit has been performed, as specified. The bill authorized a hospital located in the counties of Sacramento, San Mateo, or Santa Barbara or the city of San Jose that received the additional two-year extension to the January 2008 deadline pursuant to specified provisions to request an additional extension until September 1, 2015, to obtain either a certificate of occupancy for a replacement building or a construction final for a building on which a retrofit has been performed.	2014
AB 81	This bill authorized a hospital located in the city of Willits that received the additional two-year extension to the January 1, 2008, deadline pursuant to specified provisions to request an additional extension until September 1, 2015, to obtain either a certificate of occupancy for a replacement building or a construction final for a building on which a retrofit was performed.	2015
AB 232	This bill authorized a CAH located in the city of Tehachapi to submit a seismic safety extension application, notwithstanding specified deadlines that are earlier than the effective date of the bill and required the application to include a timetable, as specified.	2015
AB 908	This bill authorized a hospital in the Tarzana neighborhood in Los Angeles that received specified extensions to request an additional extension, as specified, until October 1, 2022, in order to obtain a certificate of occupancy from the office for a replacement building.	2017
AB 2190	The law requires OSHPD to grant an additional extension of time to a hospital that is subject to the January 1, 2020, deadline if certain conditions are met. The law authorizes the additional extension until July 1, 2022, if the compliance plan is based on replacement or retrofit, or up to five years until January 1, 2025, if the compliance plan is for a rebuild.	2019

SOURCE: Descriptions of legislation are based on OSHPD, "Changes to Seismic Safety Laws," webpage, undated.

Appendix C. History of California Earthquakes

The location, magnitude, and consequences of major earthquakes in California since the Northridge earthquake are summarized in Table C.1.

Table C.1. Significant Earthquakes in California Since 1994

Date	Name	Region	Magnitude	Deaths	Injuries	Economic Damage
8/24/2014	South Napa	North Bay	6.0 Mw	1	~200	$362M–$1B
3/28/2014	La Habra	Los Angeles area	5.1 Mw	0	Few	$10.8B
4/4/2010	Baja, California	Baja, California	7.2 Mw	2–4	100–233	$1.15B
1/9/2010	Eureka	North Coast	6.5 Mw	0	35	$21.8M–$43.0M
7/29/2008	Chino Hills	Los Angeles area	5.5 Mw	0	8	Limited
10/30/2007	Alum Rock	South Bay	5.6 Mw	0	0	Limited
12/22/2003	San Simeon	Central Coast	6.6 Mw	2	40	$250M–$300M
9/3/2000	Yountville	North Bay	5.0 Mw	0	41	$10M–$50M
10/16/1999	Hector Mine	Eastern	7.1 Mw	0	4–5	Limited
12/26/1994	Samoa	North Coast	5.5 Mw	0	0	$2.1M–$5.0M
1/17/1994	Northridge	Los Angeles area	6.7 Mw	57	8,700+	$13B–$40B

SOURCE: National Geophysical Data Center/World Data Service, Significant Earthquake Database, National Oceanic and Atmospheric Administration, undated.
NOTES: Values for deaths and injuries represent the totals experienced by the population during the event and were not attributable to damage to or within hospitals. Mw = moment magnitude scale.

References

Alesch, D., W. Petak, and L. Arendt, *Natural Hazard Mitigation Policy: Implementation, Organizational Choice and Contextual Dynamics*, New York: Springer, 2012.

Altman, Edward I., "Financial Ratios, Discriminant Analysis and the Prediction of Corporate Bankruptcy," *The Journal of Finance*, Vol. 23, No. 4, 1968, pp. 589–609.

Altman, Edward I., and Edith Hotchkiss, *Corporate Financial Distress and Bankruptcy: Predict and Avoid Bankruptcy, Analyze and Invest in Distressed Debt*, Hoboken, N.J.: Wiley, 2006.

Austin, Chris, "Earthquake Resilience of Southern California's Water Distribution Systems," Maven's Notebook, January 3, 2018. As of March 20, 2019: https://mavensnotebook.com/2018/01/03/earthquake-resilience-southern-californias-water-distribution-systems/

Babbitt, Donald H., *Improving Seismic Safety of Dams in California*, Sacramento: California Department of Water Resources, Division of Safety of Dams, 1993. As of March 20, 2019: https://water.ca.gov/-/media/DWR-Website/Web-Pages/Programs/All-Programs/Division-of-safety-of-dams/Files/Technical-Reference-docs/DSOD-Technical-References/Improving-Seismic-Safety-of-Dams-in-California---DSOD-Technical-References.pdf

Ballotpedia, "California Proposition 4, Children's Hospital Bonds Initiative (2018)," webpage, undated. As of March 20, 2019: https://ballotpedia.org/California_Proposition_4,_Children%27s_Hospital_Bonds_Initiative_(2018)

Bardach, Eugene, and Eric M. Patashnik, *A Practical Guide for Policy Analysis: The Eightfold Path to More Effective Problem Solving*, Washington, D.C.: CQ Press, 2005.

Bazzoli, G., and S. Andes, "Consequences of Hospital Financial Distress," *Hospital and Health Services Administration*, Vol. 40, No. 4, 1995, pp. 472–496.

Bland, Alastair, "How California Water Suppliers Are Getting Earthquake-Ready," KQED Science, April 18, 2018. As of March 20, 2019: https://www.kqed.org/science/1922856/how-california-water-suppliers-are-getting-earthquake-ready

Bohm, Peter, and Clifford S. Russell, "Comparative Analysis of Policy Instruments," in Allen V. Kneese and James L. Sweeney, eds., *Handbook of Natural Resource and Energy Economics*, Amsterdam: Elsevier, 1985, pp. 395–460.

Business Oregon, "Seismic Rehabilitation Grant Program," webpage, undated. As of March 20, 2019: https://www.orinfrastructure.org/Infrastructure-Programs/Seismic-Rehab/

California Assembly Bill 1823, An Act to Add and Repeal Division 20.5 (Commencing with Section 73500) of the Water Code, Relating to Regional Water Systems, California General Assembly, January 18, 2002. As of March 20, 2019: http://www.leginfo.ca.gov/pub/01-02/bill/asm/ab_1801-1850/ab_1823_bill_20020828_enrolled.html

California Assembly Bill 2190, An Act to Add Sections 130062 and 130066 to the Health and Safety Code, Relating to Hospitals, September 22, 2018. As of March 20, 2019: https://oshpd.ca.gov/wp-content/uploads/2018/10/20170AB2190_92.pdf

California Building Standards Commission, "California Building Standards Code," webpage, undated. As of March 20, 2019: https://www.dgs.ca.gov/BSC/Codes

———, *2016 California Plumbing Code: Entire Code with Errata Pages*, California Code of Regulations, Title 24, Part 5, Ontario, Calif.: International Association of Plumbing and Mechanical Officials, July 2016.

California Department of Industrial Relations, "California Consumer Price Index," webpage, undated. As of March 20, 2019: https://www.dir.ca.gov/OPRL/CAPriceIndex.htm

California Department of Transportation, "Seismic Retrofit Program," webpage, June 2014. As of March 20, 2019: http://www.dot.ca.gov/hq/paffairs/about/retrofit.htm

California Department of Water Resources, "Lake Perris and Perris Dam Projects," webpage, undated. As of March 20, 2019: https://water.ca.gov/Programs/Engineering-And-Construction/Perris-Dam-Remediation

California Energy Commission, "2019 Building Energy Efficiency Standards," webpage, undated. As of March 20, 2019: https://www.energy.ca.gov/title24/2019standards/

California Planning and Development Report, "Cities Pressure San Francisco to Repair Hetch Hetchy," April 1, 2002. As of March 20, 2019: http://www.cp-dr.com/articles/node-951

California Pollution Control Financing Authority, "California Capital Access Program (CalCAP) Seismic Safety Financing Program," webpage, undated. As of March 20, 2019: https://www.treasurer.ca.gov/cpcfa/calcap/seismic/summary.asp

California Residential Mitigation Program, Earthquake Brace and Bolt Program, homepage, undated. As of March 20, 2019: https://www.earthquakebracebolt.com

California Seismic Safety Commission, *Northridge Earthquake: Turning Loss to Gain*, Sacramento, 1995. As of March 20, 2019: https://archive.org/stream/804210-1995-state-seismic-safety-report/804210-1995-state-seismic-safety-report_djvu.txt

———, *Progress Report for the California Earthquake Loss Reduction Plan: 2002–2006*, Sacramento, December 2003. As of March 20, 2019: https://ssc.ca.gov/forms_pubs/cssc_2003-06_progressreport_ca_elrp.pdf

California Senate Bill 90, An Act to Amend Section 130060 of the Health and Safety Code, and to Amend Sections 14105.281, 14166.115, and 14167.10 of, to Amend and Repeal Section 14166.245 of, and to Add and Repeal Article 5.226 (Commencing with Section 14168.1) and Article 5.227 (Commencing with Section 14168.31) of Chapter 7 of Part 3 of Division 9 of, the Welfare and Institutions Code, Relating to Health, Making an Appropriation Therefor, and Declaring the Urgency Thereof, to Take Effect Immediately, April 13, 2011. As of March 20, 2019: https://oshpd.ca.gov/wp-content/uploads/2018/06/sb_90_bill_20110413_chaptered.pdf

California Senate Bill 306, An Act to Amend Section 129765 of, and to Add Section 130061.5 to, the Health and Safety Code, Relating to Health Facilities, October 13, 2007. As of March 20, 2019: https://oshpd.ca.gov/wp-content/uploads/2018/06/sb_306_bill_20071013_chaptered.pdf

California Senate Bill 519, Alfred E. Alquist Hospital Facilities Seismic Safety Act of 1973, 1973.

California Senate Bill 1801, An Act to Amend Section 130060 of the Health and Safety Code, Relating to Health Facilities, September 29, 2000. As of March 20, 2019: https://oshpd.ca.gov/wp-content/uploads/2018/06/SB1801.pdf

California Senate Bill 1953, An Act to Amend Section 18938 of, and to Add Article 8 (Commencing with Section 15097.100) and Article 9 (Commencing with Section 15097.125) to Chapter 1 of Division 12.5 of, the Health and Safety Code, Relating to Building Standards, and Making an Appropriation Therefor, September 21, 1994. As of March 20, 2019: http://www.leginfo.ca.gov/pub/93-94/bill/sen/sb_1951-2000/sb_1953_bill_940922_chaptered

California Senate Health Committee, bill analysis of Senate Bill SB 499, Sacramento, April 15, 2009. As of March 20, 2019: http://www.leginfo.ca.gov/pub/09-10/bill/sen/sb_0451-0500/sb_499_cfa_20090413_154541_sen_comm.html

California Senate Rules Committee, bill analysis and third reading of Senate Bill 1661, Sacramento, April 17, 2006. As of March 20, 2019: http://www.leginfo.ca.gov/pub/05-06/bill/sen/sb_1651-1700/sb_1661_cfa_20060510_123423_sen_floor.html

California's Office of Statewide Health Planning and Development, "Application Process for the Seismic Compliance Unit," webpage, undated. As of March 20, 2019: https://oshpd.ca.gov/construction-finance/seismic-compliance-and-safety/seismic-application-process/

———, "Changes to Seismic Safety Laws," webpage, undated. As of March 20, 2019: https://oshpd.ca.gov/construction-finance/seismic-compliance-and-safety/seismic-legislation/

———, "Construction Financing: Cal-Mortgage Loan Insurance Program," webpage, undated. As of March 20, 2019: https://oshpd.ca.gov/construction-finance/construction-financing

———, "Hospital Financials," webpage, undated. As of September 15, 2018: https://oshpd.ca.gov/data-and-reports/cost-transparency/hospital-financials/

———, "Plan Review Processes and Goals," webpage, undated. As of February 8, 2019: https://oshpd.ca.gov/construction-finance/building-and-construction-projects/plan-review-processes-goals/

———, "Seismic Performance Ratings," webpage, undated. As of March 20, 2019: https://oshpd.ca.gov/construction-finance/seismic-compliance-and-safety/seismic-performance-ratings/

———, "Seismic Compliance Program Overview," Sacramento, 2015. As of March 20, 2019: https://www.dgs.ca.gov/-/media/Divisions/BSC/03-Rulemaking/18MonthCodeAdoption/2015-Triennial-Cycle/2015-Public-Comment/11-23-15-GRN-PEME-45-day/OSHPD-05-15/Exhibit54.ashx?la=en&hash=6BEE0DC0BABCFAAD26AA838225242F04D6AE0810 https://oshpd.ca.gov/construction-finance/seismic-compliance-and-safety/program-overview/

———, *Cal-Mortgage Loan Insurance Division Monthly Activity Report Ending November 30, 2018*, Sacramento, 2018. As of March 20, 2019: https://oshpd.ca.gov/ml/v1/resources/document?rs:path=/Construction-And-Finance/Documents/Construction-Financing/Reports/Monthly/2018NovemberMonthlyReport.pdf

———, "Hospital Building Data," data set, accessed September 2018. As of March 20, 2019: https://data.chhs.ca.gov/dataset/hospital-building-data

California Transportation Commission, "Adoption of Proposition 1B Local Bridge Seismic Retrofit Guidelines," Resolution LBS1B-G-0708-01, May 29, 2008. As of March 20, 2019: http://www.catc.ca.gov/programs/prop1b/lbsra/docs/LBSRA_guidelines_052908.pdf

Capps, C., D. Dranove, and C. Ody, "The Effect of Hospital Acquisitions of Physician Practices on Prices and Spending," *Journal of Health Economics*, Vol. 59, 2018, pp. 139–152.

Chang, Tom, and Mireille Jacobson, "Natural Disaster Management: Earthquake Risk and Hospitals' Provision of Essential Services in California," in Daniel P. Kessler, ed., *Regulation Versus Litigation: Perspectives from Economics and Law*, Chicago: University of Chicago Press, 2010, pp. 301–320.

———, "What Do Nonprofit Hospitals Maximize? Evidence from California's Seismic Retrofit Mandate," unpublished manuscript, 2012. As of March 20, 2019: https://users.nber.org/~changt/ChangJacobson13apr9.pdf

Cheevers, J., and A. Abrhamson, "Earthquake: The Long Road Back; Hospitals Strained to the Limit by Injured; Medical Care; Doctors Treat Quake Victims in Parking Lots; Details of Some Disaster-Related Deaths Are Released," *Los Angeles Times*, January 19, 1994. As of March 20, 2019: http://articles.latimes.com/1994-01-19/local/me-13343_1_parking-lot

Coleman, Paul A., *2019 California Building Standards Code Update for the California Hospital Association*, Sacramento, Calif.: Office of Statewide Health Planning and Development, 2018. As of March 20, 2019: https://www.calhospital.org/sites/main/files/file-attachments/2019_cbsc_update_for_cha_0.pdf

Dineen, J. K., "For SF Landlords, Deadline Looms to Retrofit Soft-Story Buildings," *San Francisco Chronicle*, April 9, 2017. As of March 20, 2019: https://www.sfchronicle.com/bayarea/article/For-SF-landlords-deadline-looms-to-retrofit-11060397.php

DOT—*See* California Department of Transportation.

Ellis, Virginia, "1,364 Additional Bridges Are Put on Retrofit List: Earthquakes; Caltrans Report Says 302 Highway Structures Needing Seismic Strengthening Are in L.A. County; Plans Call for an Accelerated Schedule That Would Be Completed by 1997," *Los Angeles Times*, July 1, 1994. As of March 20, 2018: http://articles.latimes.com/1994-07-01/news/mn-10784_1_highway-structures

Federal Emergency Management Agency, *Seismic Evaluation and Retrofit of Multi-Unit Wood-Frame Buildings with Weak First Stories*, Washington, D.C., May 2012. As of March 20, 2019: https://www.fema.gov/media-library-data/20130726-1916-25045-2624/femap_807.pdf

———, "Earthquake Grants," webpage, last updated June 14, 2018. As of March 20, 2019: https://www.fema.gov/earthquake-grants

FEMA—*See* Federal Emergency Management Agency.

Flannery, T., and D. Hofrichter, "What's the Right Incentive Compensation Plan?" *Trustee: The Journal for Hospital Governing Boards*, Vol. 60, No. 7, 2007, pp. 22–26.

Fracassa, Dominic, "$823 Million, 31 Billion-Gallon Calaveras Reservoir Dam Ready for Debut," *San Francisco Chronicle*, September 21, 2018. As of March 20, 2019: https://www.sfchronicle.com/bayarea/article/Crews-topping-off-Calaveras-Dam-shoring-up-Bay-13245983.php

Gillengerten, J., "Estimated Cost of Structural Retrofits," paper presented at the SB 1953 Seismic Compliance Meeting, California Healthcare Association, Sacramento, March 30, 1999.

Goldman, D. P., M. Vaiana, and J. A. Romley, "The Emerging Importance of Patient Amenities in Hospital Care," *New England Journal of Medicine*, Vol. 363, No. 23, 2010, pp. 2185–2187.

Harper, Esperanza, Nicholas Watkins, and Tamra Minnier, "Patient Room Size Affects Satisfaction," Design Works HC, May 16, 2014.

Healthcare Financial Management Association, "Key Hospital Financial Statistics and Ratio Medians: Glossary of Formulas," webpage, undated. As of March 20, 2019: https://www.hfma.org/Content.aspx?id=1113

Hillman, Anne, "Emergency Rooms Open, Non-Critical Appointments Canceled at Anchorage Area Hospitals," Alaska Public Media, November 30, 2018. As of March 20, 2019: https://www.alaskapublic.org/2018/11/30/emergency-rooms-open-non-critical-appointments-canceled-at-anchorage-area-hospitals/

Holmes, G. M., B. G. Kaufman, and G. H. Pink, "Predicting Financial Distress and Closure in Rural Hospitals," *The Journal of Rural Health*, Vol. 33, No. 3, 2017, pp. 239–249.

Internal Revenue Service, "Community Health Needs Assessment for Charitable Hospital Organizations—Section 501(r)(3)," webpage, last updated November 7, 2018. As of March 20, 2019: https://www.irs.gov/charities-non-profits/community-health-needs-assessment-for-charitable-hospital-organizations-section-501r3

Joffe, Marc, "California Healthcare Districts in Crisis," California Policy Center, January 22, 2015. As of March 20, 2019: https://californiapolicycenter.org/california-healthcare-districts-in-crisis/

KPCC, "The Northridge Earthquake, 20 Years Later," Southern California Public Radio, 2004. As of March 20, 2019: https://www.scpr.org/northridge-earthquake

Langabeer, J., "Predicting Financial Distress in Teaching Hospitals," *Journal of Health Care Finance*, Vol. 33, No. 2, 2006, pp. 84–92.

Langabeer, James R., II, Karima H. Lalani, Tiffany Champagne-Langabeer, and Jeffrey R. Helton, "Predicting Financial Distress in Acute Care Hospitals," *Hospital Topics*, Vol. 96, No. 3, 2018.

LePointe, Jacqueline, "How Hospital Merger and Acquisition Activity Is Changing Healthcare," Revcycle Intelligence, July 20, 2018. As of March 20, 2019: https://revcycleintelligence.com/features/how-hospital-merger-and-acquisition-activity-is-changing-healthcare

Lin, Rong-Gong, II, "Big Earthquake Would Topple Countless Buildings, but Many Cities Ignore the Danger," *Los Angeles Times*, January 17, 2019. As of March 20, 2019: https://www.latimes.com/local/lanow/la-me-northridge-earthquake-california-20190117-story.html

Liu, L. L., K. J. Jervis, M. M. Z. Younis, and D. A. Forgione, "Hospital Financial Distress, Recovery and Closure: Managerial Incentives and Political Costs," *Journal of Public Budgeting, Accounting and Financial Management*, Vol. 23, No. 1, 2011, pp. 31–68.

Mallett, William J., Nicole T. Carter, and Peter Folger, *Earthquake Risk and U.S. Highway Infrastructure: Frequently Asked Questions*, Washington, D.C.: Congressional Research Service, May 2016. As of March 20, 2019: https://fas.org/sgp/crs/misc/R41746.pdf

Mayoral Seismic Safety Task Force, *Resilience by Design*, Los Angeles, Calif., December 2014. As of March 20, 2019: https://d3n8a8pro7vhmx.cloudfront.net/mayorofla/pages/16797/attachments/original/1418064797/Earthquake_report_FINAL_Dec_8_full_report_compressed.pdf?1418064797

McCue, M. J., and J. M. Thompson, "Hospital Compliance with a State Unfunded Mandate: The Case of California's Earthquake Safety Law," *Hospital Topics*, Vol. 90, No. 4, 2012, pp. 91–97.

Meade, Charles, and Jonathan Kulick, *SB1953 and the Challenge of Hospital Seismic Safety in California*, Oakland: California HealthCare Foundation, 2007. As of March 20, 2019: https://www.rand.org/pubs/external_publications/EP20061224.html

Meade, Charles, Jonathan Kulick, and Richard Hillestad, *Estimating the Compliance Costs for California SB 1953*, Oakland: California HealthCare Foundation, 2002. As of March 20, 2019: https://www.chcf.org/wp-content/uploads/2017/12/PDF-ComplianceCostsForSB1953.pdf

Metropolitan Water District of Southern California, *Modernizing the System: California Waterfix Finance and Cost Allocation*, Los Angeles, 2017. As of March 20, 2019: http://www.mwdh2o.com/docsvcspubs/waterfix/assets/cawaterfix_finance_costallocation_whitepaper_factsheet.pdf

National Fire Protection Association, *NFPA 101, Life Safety Code*, Quincy, Mass., 2018. As of March 20, 2019:
https://downloads.nfpa.org/codes-and-standards/all-codes-and-standards/list-of-codes-and-standards/detail?code=101

National Geophysical Data Center/World Data Service, Significant Earthquake Database, National Oceanic and Atmospheric Administration, undated. As of April 1, 2019:
https://www.ngdc.noaa.gov/nndc/struts/form?t=101650&s=1&d=1

Occupational Safety and Health Administration, "Law and Regulations," webpage, undated. As of March 20, 2019:
https://www.osha.gov/law-regs.html

Oregon Seismic Safety Policy Advisory Commission, *The Oregon Resilience Plan: Reducing Risk and Improving Recovery for the Next Cascadia Earthquake and Tsunami*, Salem, February 2013.

OSHPD—*See* California's Office of Statewide Health Planning and Development.

Pao, Jean Lin, Michael Siewart, Jeremy Creelan, and Henry Mere, *Preparing for the "Big One": Saving Lives Through Earthquake Mitigation in Los Angeles, California*, Washington, D.C.: U.S. Department of Housing and Urban Development, 1995.

Poland, C., "California's Wise Retreat from Seismic Overreaction," *Healthcare Design*, May 31, 2008. As of March 20, 2019:
https://www.healthcaredesignmagazine.com/architecture/californias-wise-retreat-seismic-overreaction/

Public Law 111-148, Patient Protection and Affordable Care Act, March 10, 2010. As of December 27, 2018:
https://www.govinfo.gov/app/details/PLAW-111publ148

Ramamonjiarivelo, Z. H., "Is Privatization the Solution to the Financial Distress of Public Hospitals?" *Academy of Management Proceedings*, Vol. 2014, No. 1, 2014.

Rider Levett Bucknall, "Publications," webpage, undated. As of March 20, 2019:
https://www.rlb.com/en/index/publications/?filter-publication-type=Research&filter-region=americas

———, *North America Quarterly Construction Cost Report: Fourth Quarter 2017*, December 2017. As of March 20, 2019:
https://www.rlb.com/wp-content/uploads/2018/01/Q4-2017-QCR.pdf

———, *North America Quarterly Construction Cost Report: Fourth Quarter 2018*, December 2018. As of March 20, 2019:
https://www.rlb.com/wp-content/uploads/2019/01/Q4-QCR-2018.pdf

San Francisco Department of Building Inspection, "Soft Story Properties List," webpage, undated. As of March 14, 2019: https://sfdbi.org/soft-story-properties-list

Santa Monica Rent Control Board, "Study Session on Consideration of Tenant Protections and Costs Associated with the City of Santa Monica's Mandatory Seismic Retrofit Program," letter from Tracy Condon to the Rent Control Board commissioners, October 12, 2017. As of March 20, 2019: https://www.smgov.net/uploadedFiles/Departments/Rent_Control/About_the_Rent_Control_Board/Staff_Reports/2017/Item%2012A%20%20Seismic%20Retrofitting%20Study%20Session.pdf

Santos-Burgoa, Carlos, Ann Goldman, Elizabeth Andrade, Nicole Barre, Uriyoan Colon-Ramos, Mark Edberg, Alejandra Garcia-Meza, Lynn Goldman, Amira Roess, John Sandberg, and Sco Zeger, *Ascertainment of the Estimated Excess Mortality from Hurricane Maria in Puerto Rico*, Washington, D.C.: George Washington University, Health Sciences Research Commons, 2018.

Scheffler, Richard, *California's Closed Hospitals, 1995–2000*, Berkeley: Nicholas C. Petris Center on Health Care Markets and Consumer Welfare, University of California, Berkeley School of Public Health, 2001. As of March 20, 2019: http://petris.org/wp-content/uploads/2013/02/ClosedHospitals.pdf

Schultz, C. H., K. L. Koenig, and R. J. Lewis, "Implications of Hospital Evacuation After the Northridge, California, Earthquake," *New England Journal of Medicine*, Vol. 348, No. 14, 2003, pp. 1349–1355.

Seismic Ordinances of California, "Los Angeles: Non-Ductile Concrete Structures," webpage, undated. As of March 20, 2019: http://www.seismicordinances.com/non-ductile-concrete-structures/los-angeles

State of California, California Health and Safety Code, Division 107: Statewide Health Planning and Development, Part 7: Facilities Design Review and Construction, 2019.

Tierney, Kathleen J., "Social Impacts and Emergency Response," in John F. Hall, ed., *Northridge Earthquake, January 17, 1994: Preliminary Reconnaissance Report*, Oakland, Calif.: Earthquake Engineering Institute, 1994, pp. 87–93. As of March 20, 2019: http://www.desastres.hn/docum/crid/Septiembre-Octubre2005/CD-1/pdf/eng/doc6099/doc6099-contenido.pdf

U.S. Department of Health and Human Services, "Hospital Preparedness Program (HPP)," webpage, last reviewed March 4, 2019. As of March 20, 2019: https://www.phe.gov/Preparedness/planning/hpp/Pages/default.aspx

U.S. Department of Justice, *2010 ADA Standards for Accessible Design*, Washington, D.C., September 15, 2010. As of March 20, 2019: https://www.ada.gov/regs2010/2010ADAStandards/2010ADAstandards.htm

Xia, Rosana, and Rong-Gong Lin II, "Earthquake Could Destroy L.A.'s Water Lifeline," *Los Angeles Times*, December 15, 2014. As of March 20, 2019: https://www.latimes.com/local/california/la-me-quake-water-20141216-story.html

Yamauchi, Hirotaka, "Characteristics and Cost Sharing of Transport Infrastructure," *JBIC Review*, No. 3, December 2000, pp. 41–51.

Yune, Howard, "California Rates Napa's Lake Hennessey Dam in 'Fair' Condition," *Napa Valley Register*, September 5, 2017. As of March 20, 2019: https://napavalleyregister.com/news/local/california-rates-napa-s-lake-hennessey-dam-in-fair-condition/article_afd2d5de-505a-5b87-ad10-950e3c200883.html